ELEPHANTS ON ACID

Alex Boese holds a master's degree in the history of science from the University of California, San Diego. He is the author of *The Museum of Hoaxes* and *Hippo Eats Dwarf* and the creator of www.museumofhoaxes.com. He lives near San Diego.

Also by Alex Boese

THE MUSEUM OF HOAXES

HIPPO EATS DWARF
A Field Guide to Hoaxes and Other B.S.

Alex Boese

Elephants on Acid

and Other Bizarre Experiments

PAN BOOKS

First published 2007 by Harvest, an imprint of Harcourt, Inc., Orlando, Florida

First published in Great Britain 2008 by Boxtree, an imprint of Pan Macmillan Ltd

This edition published 2009 by Pan Books
an imprint of Pan Macmillan Ltd
Pan Macmillan, 20 New Wharf Road, London N1 9RR
Basingstoke and Oxford
Associated companies throughout the world
www.panmacmillan.com

ISBN 978-0-330-50664-9

9 8 7 6 5 4 3 2 1

A CIP catalogue record for this book is available from
the British Library.

Printed in the UK by CPI Mackays, Chatham ME5 8TD

Visit www.panmacmillan.com to read more about all our books
and to buy them. You will also find features, author interviews and
news of any author events, and you can sign up for e-newsletters
so that you're always first to hear about our new releases.

Once again, to Beverley

CONTENTS

INTRODUCTION

In the following pages you will encounter elephants on LSD, two-headed dogs, zombie kittens, and racing cockroaches—to name just a few of the oddities that await you. Some of these oddities might shock you. Others might amuse you. Still others might make you think, "That can't be true!" However, I assure you, unless stated otherwise, it's all true. This is definitely a work of nonfiction.

All of these strange phenomena share one thing in common: They have all played starring roles in scientific experiments. What you're holding in your hands is a collection of the most bizarre experiments ever conducted. No knowledge of science is needed to appreciate them, just curiosity and an appreciation for the odd.

The criteria for inclusion: Did an experiment make me chuckle, shake my head in disbelief, grimace with disgust, roll my eyes, or utter a shocked exclamation? Did it force me to wonder what kind of imagination, twisted or brilliant, could have dreamed up such a thing? If so, it went on the *must include* pile. As for the question of scientific worth, some of these experiments are brilliant examples of the scientific method; others are not. Mad scientists, geniuses, heroes, villains, and fools all rub shoulders here.

I first encountered the bizarre-experiment genre in the mid-1990s as a graduate student studying the history of science at the University of California, San Diego. My formal studies focused on all the usual suspects—Darwin, Galileo,

Newton, Copernicus, Einstein, et al. But scattered through-out the texts assigned by my professors were references to little-known, intriguing tales about crackpots and mad experi-menters. These secondary tales were far more interesting to me than the primary material I was supposed to be learning. Soon I found myself in the library chasing down those stories.

Fast-forward to 2005. I had built a kind of career—*kind of* because my friends and family insist what I do is too much fun to be a real job—out of studying another offbeat subject I encountered during the seven years I spent at grad school. That subject was hoaxes. Think Orson Welles's 1938 *War of the Worlds* broadcast or the Piltdown Man. I created a Web site about hoaxes, museumofhoaxes.com, and authored two books on the topic.

One day I was having lunch with my American editor, Stacia Decker. As we ate our meals, she told me about an unusual experiment involving a researcher who raced cock-roaches. She had heard the story from her sister. Apparently, a scientist had built a little stadium, complete with stands in which other roaches could sit to watch the races. (You can read more about the roach stadium in chapter five.) Bizarre experiments would make a pretty good topic for a book, she suggested. It would, I agreed, as I thought back to all the material I had encountered in graduate school. The book you're reading now is the result of that conversation.

Shifting from hoaxes to bizarre experiments continued my interest in weird stuff. But I also came to realize that hoaxes and bizarre experiments share many features in common.

An experiment starts when a researcher looks at a situation and thinks, *What would happen if I changed one part of this?* He or she performs an experimental manipulation and observes the results. A hoax proceeds in essentially the same way, except that the manipulation takes the form of an outrageous lie. Of course, as we'll see throughout this book, the manipulations performed by researchers also frequently involve deception. Experimenters sometimes rehearse for days, perfecting the elaborate ruses they're going to foist on their unsuspecting subjects. In these cases, the line separating hoaxes and experiments is almost indistinguishable.

The big difference between hoaxes and bizarre experiments is that experimenters wrap themselves in the authority of science. They claim as their motive the desire to advance knowledge, whereas hoaxers are often just trying to get a laugh or perpetrate a scam. This sense of gravity is what lends bizarre experiments their particularly surreal quality. It's that odd combination of apparent seriousness—white-lab-coat-wearing researchers toiling dispassionately to further the limits of knowledge—mixed with a hint of mischief, eccentricity, or, in some cases, seeming insanity, that provides the frisson of weirdness. To preserve this effect, I've avoided including any experiments conducted in a spirit of jest. All the research in the following pages was undertaken quite seriously. To me, this makes these stories all the more fascinating.

Let me wrap up these introductory remarks by addressing a few questions that may occur to you as you read this book:

Hey, Where are the Nazis?

I wouldn't mention this, except that the Nazi death-camp experiments are apparently what many people think of first when the subject of bizarre experiments comes up. At least, whenever I told people I was writing a book about bizarre experiments, the most common response I received was, "You mean, like the Nazi experiments?"

I have not included any Nazi research in this book. First, because I didn't intend the book to be a catalog of atrocities. Second, because I wanted to explore actual scientific research —not sadistic torture disguised as science, which is what I consider the Nazi "experiments" to be.

How can one distinguish between the two? A couple of guidelines suggest themselves. First, once an experimenter starts purposefully killing people, his research instantly ceases to be legitimate. The second rule is more subtle: Genuine scientists publish their work. When a researcher submits his work for publication, he offers it up to the scrutiny of the scientific community. And when an established, respected journal accepts the submission, this suggests other scientists agree it deserves wider dissemination and consideration. It doesn't mean the work is good science, or ethically justified —especially when judged by present-day standards. But it does mean that, for better or worse, the research cannot be denied a place in the history of science. Sometimes extenuating circumstances prevent a researcher from publishing his work, but 99 percent of the time, the publication rule is a useful guideline for identifying real science.

Where's my favorite bizarre experiment?

Maybe there's a bizarre experiment that's a particular favorite of yours, and you discover that—uh-oh—it isn't in here. It could happen. The book format does not permit unlimited space. Forced to pick and choose from a wide field of possibilities, I ultimately settled on ten themes, each of which became the focus of a chapter. If an experiment didn't relate to one of these themes, I put it aside.

How can I find out more about an experiment?

I don't dwell too long on any one subject. If all went as planned, this should make the book fast-paced and easy to read. I hope that people who wouldn't normally read a book about science might enjoy these stories. I joke that it's a toilet reader's guide to science—which is why I have included chapter eight specifically for this audience.

This format means that each vignette presents a condensed account of what is often a very complex subject. I've placed a single reference at the end of each vignette. This reminds you that the story you just read is real. I wasn't making it up. But I've also provided additional references at the end of the book so that readers can pursue in greater depth any topic that whets their interest.

One more comment, then I'll let you get on to the good stuff—the experiments.

Although this book may, at first glance, resemble a kind of circus parade of oddities (led by an elephant on acid, no less), my intention is *not* to trivialize scientific research or the experimenters who appear in the following pages. Quite the opposite. To me, what these stories are really about is people consumed by insatiable curiosity.

The researchers who appear in the following pages—even the scariest and most eccentric ones—all share one virtue. They all looked at the world around them, and instead of taking what they saw for granted, they asked questions. Their questions might have been bizarre. They might even have been stupid. But often the most brilliant discoveries come from people willing to ask what might seem, at the time, to be dumb questions.

The danger of curiosity is that only in hindsight do people know whether it's led them to brilliance or madness, or somewhere in between. Once you fall under its spell, you're along for the ride, wherever it may take you.

Like the researchers I was writing about, I, too, experienced a kind of obsessive curiosity as I worked on this book. I spent months in the library, pulling dusty old journals down from shelves, eagerly flipping from one page to the next, always looking for something new that would catch my eye. The other library patrons must have wondered who *was* that odd man, chuckling as he read decades-old copies of the *Journal of Personality and Social Psychology*. Hopefully you'll find these experiments as fascinating to read about as I found them to write about.

—Alex
April 2007

CHAPTER ONE

Frankenstein's Lab

Beakers bubble over. Electricity crackles. A man hunches over a laboratory bench, a crazed look in his eyes. This is the classic image of a mad scientist—a pale-skinned, sleep-deprived man toiling away in a lab full of strange machinery, delving into nature's most forbidden and dreadful secrets. In the popular imagination, no one embodies this image better than Victor Frankenstein, the titular character of Mary Shelley's 1818 novel. Gathering material from charnel houses and graves, he created an abomination—a living monster pieced together from the body parts of the dead. But he was just fictional, right? Surely no one has done that kind of stuff in real life. Well, perhaps no one has *succeeded* in creating an undead monster, but it hasn't been for lack of trying. The history of science is full of researchers whose experiments have, like Frankenstein's, gone well beyond conventional boundaries of morality and plunged them deep into the realms of the morbid and bizarre. These are the men—for some reason, they are all men—we meet in this chapter. Prepare yourself for zombie kittens, two-headed dogs, and other lab-spawned monstrosities.

The Body Electric

"Frog soup," Madame Galvani wheezed. "Make me some frog soup." She had been sick in bed for over a week, aching, feverish, and suffering from a wracking cough. The doctor had diagnosed consumption. Frog soup, he assured her, was just the thing to put her on the road to recovery. She asked her servants to prepare some, and soon they were scurrying about, gathering the ingredients. Painfully, she forced herself out of bed to supervise. It was just as well she did so. She found them milling around, searching for somewhere to lay out the frogs. "Put them on the table in my husband's lab," Madame Galvani instructed. A servant obediently carried the tray of skinned frogs into the lab and set it down next to one of the doctor's electrical machines. He picked up a knife and began to carve a frog, but just then a spark flew from the machine and touched the knife. Instantly the legs of the frog twitched and spasmed. Madame Galvani, who had followed the servant in, gasped in surprise. "Luigi, come quick," she cried. "The most remarkable thing has just happened."

In 1780 Luigi Galvani, an Italian professor of anatomy, discovered that a spark of electricity could cause the limbs of a dead frog to move. Nineteenth-century popularizers of science would later attribute this discovery to his wife's desire for frog soup. Unfortunately, that part of the story is a legend. The reality is that Galvani was quite purposefully studying frogs, to understand how their muscles contracted, when a spark caused movement in a limb. However, the frog-soup story does have the virtue of restoring to his wife a greater role in the discovery than Galvani granted her—credit she probably deserves since she was a highly educated woman from a family of scientists. And Madame Galvani did develop consumption, and may well have been treated with frog soup. Unfortunately, the frog soup didn't help her. She died in 1790.

A year after his wife's death, Galvani finally published an account of the experiment. It caused a sensation throughout Europe. Many believed Galvani had discovered the hidden secret of life. Other men of science rushed to repeat the experiment, but it didn't take them long to grow bored with frogs and turn their attention to more interesting animals. *What would happen,* they wondered, *if you wired up a human corpse?*

Galvani's nephew, Giovanni Aldini, took the initiative and pioneered the art of corpse reanimation. He promoted his publicity-shy uncle's work by embarking on a tour of Europe in which he offered audiences the greatest (or, at least, most stomach-wrenching) show they'd ever seen—the electrification of a human body.

Aldini's most celebrated demonstration occurred in London on January 17, 1803, before an audience of the Royal College of Surgeons. The body of twenty-six-year-old George Forster, executed for the murder of his wife and child, was whisked straight from the gallows to Aldini and his waiting crowd. Aldini then attached parts of Forster's body to the poles of a 120-plate copper-and-zinc battery.

First the face. Aldini placed wires on the mouth and ear. The jaw muscles quivered, and the murderer's features twisted in a rictus of pain. The left eye opened as if to gaze upon his torturer. Aldini played the body like a marionette, moving wires from one body part to another, making the back arch, the arms beat the table, and the lungs breathe in and out.

For the grand finale he hooked one wire to the ear and plunged the other up the rectum. Forster's corpse broke into a hideous dance. The *London Times* wrote of the scene: "The right hand was raised and clenched, and the legs and thighs were set in motion. It appeared to the uninformed part of the bystanders as if the wretched man was on the eve of being restored to life."

A few days later Aldini continued his London tour with a show at a Dr. Pearson's lecture room. There he unveiled the decapitated head of an ox and extended its tongue out of its mouth by means of a hook. Then he turned on the current. The tongue retracted so rapidly that it tore itself off the hook, while simultaneously "a loud noise issued from the mouth by the absorption of air, attended by violent contortions of the whole head and eyes." Science had at last created an electric belching ox head.

An even more spectacular demonstration occurred on November 4, 1818, in Glasgow, when Scottish chemist (and later industrial capitalist) Andrew Ure connected the corpse of the executed murderer Matthew Clydesdale to a massive 270-plate battery. *Twice the power, twice the fun.* When he linked the spinal marrow to the sciatic nerve, "every muscle in the body was immediately agitated with convulsive movements, resembling a violent shuddering from cold." Connecting the phrenic nerve to the diaphragm provoked "full, nay, laborious breathing . . . The chest heaved, and fell; the belly was protruded, and again collapsed, with the relaxing and retiring diaphragm." Finally Ure joined the poles of the battery to an exposed nerve in the forehead and to the heel: "Every muscle in his countenance was simultaneously thrown into fearful action; rage, horror, despair, anguish, and ghastly smiles, united their hideous expression in the murderer's face, surpassing far the wildest representations of a Fuseli or a Kean." Some spectators fainted, and others fled the lecture hall in terror.

Men of science such as Aldini and Ure were confident galvanic electricity could do far more than provide a macabre puppet show. They promised that, under the right circumstances, it could restore life itself. Ure wrote of his experiment on the murderer Clydesdale, "There is a probability that life might have been restored. This event, however little desirable with a murderer, and perhaps contrary to law, would yet have been pardonable in one instance, as it would have been highly honourable and useful to science."

As late as the 1840s, English physicist William Sturgeon (inventor of the first electromagnets) described electrifying the bodies of four drowned young men in an attempt to bring them back to life. He failed but felt sure he would have succeeded had he only reached the scene sooner.

Mary Shelley never indicated on whom she had based her character of Victor Frankenstein, but the experimental electrification of corpses was undeniably a source of inspiration for her. In the introduction to the 1831 edition of *Frankenstein*, she wrote that the idea for the novel came to her in June 1816, after she overheard Lord Byron and Percy Shelley discussing recent galvanic experiments and speculating about the possibility that electricity could restore life to inanimate matter. That night she had a nightmare about a "pale student of unhallowed arts kneeling beside the thing he had put together. I saw the hideous phantasm of a man stretched out, and then, on the working of some powerful engine, show signs of life, and stir with an uneasy, half-vital motion." And so, from a journey of discovery that began with a twitching frog, Victor Frankenstein and his monster were born.

Aldini, G. (1803). *An account of the galvanic experiments performed by John Aldini, . . . on the body of a malefactor executed at Newgate, Jan. 17, 1803. With a short view of some experiments which will be described in the author's new work now in the press.* London: Cuthell and Martin.

Zombie Kitten

During the early nineteenth century many researchers repeated the galvanic experiment of electrifying a corpse. But only one man claimed to have used the technique to restore life to the dead. His name was Karl August Weinhold.

Weinhold published a work, *Experiments on Life and its Primary Forces through the Use of Experimental Physiology*, in which he detailed an experiment that, supposedly, succeeded in revivifying a decapitated kitten.

The procedure went as follows. First, he took a three-week-old kitten and removed its head. Next, he extracted the spinal cord and completely emptied the hollow of the spinal column with a sponge attached to a screw probe. Finally, he filled the hollow with an amalgam of silver and zinc. The metals acted as a battery, generating an electric current that immediately brought the kitten to life—so he said. Its heart revived, and for a few minutes it pranced and hopped around the room. Weinhold wrote, "Hopping around was once again stimulated after the opening in the spinal column was closed. The animal jumped strongly before it completely wore down." To modern readers, his creation may sound disturbingly like a mutant version of the Energizer Bunny.

Historians believe that Weinhold performed this experiment, but the consensus is that he lied about the results. After all, a kitten lacking a brain and spinal column is not going to dance around a room, no matter how much electricity you pump into it. As medical historian Max Neuburger delicately put it, "His experiments illustrate the fantasy of his thinking and observations."

Weinhold probably would have preferred to use a human corpse instead of a kitten, but in 1804 German authorities had banned the further use of human bodies in galvanic experiments. The public, it seemed, had lost its stomach for such postmortem grotesqueries. Thus restricted, Weinhold focused his efforts on animals. He was willing to break the laws of nature, but not of the German state.

Weinhold's personal life matched the strangeness of his experiment. His contemporaries described him as peculiarly unattractive. His long arms and legs contrasted with his small head, and his voice sounded feminine. He had no beard. He made many enemies on account of his campaign to eliminate poverty by forcibly infibulating indigent men—*infibulate* meaning to sew the foreskin shut. Whether this crusade was in any way inspired by the deformity of his own genitals, a condition discovered by a medical examiner after his death, is not known. A modern biographer of his noted, "Weinhold seems to have cared little for what others thought about him, and he was not afraid to propose ideas that would cause large segments of the population to despise or detest him."

If ever there was a real-life Dr. Frankenstein, it was Weinhold. But did he actually serve as a model for Shelley's character? Historians have speculated about this possibility, but it is unlikely. For one thing, Weinhold published his work in 1817, a year after Shelley began work on her novel.

Perhaps horror fans should be thankful that Shelley wasn't

aware of Weinhold. Otherwise she might have been tempted to change her novel to fit his story. Imagine a mob of villagers armed with pitchforks and torches chasing after a headless zombie kitten. It just wouldn't have been the same.

Weinhold, K. A. (1817). *Versuche über das Leben und seine, Grundkräfte, auf dem Wege der experimental-Physiologie*. Magdeburg: Creutz.

The Electrical Acari

"Life! I have created life!" Andrew Crosse gazed down at the small white insects crawling in the liquid-filled basin. Then he threw back his head and laughed maniacally.

In a Hollywood version of history, that would have been Crosse's reaction to the unusual discovery he made in 1836. But in real life his reaction was probably more along the lines of, "I say, how astonishing."

Crosse was a Victorian gentleman who lived in a secluded mansion in rural Somerset, England. From an early age he had been fascinated by electrical phenomena, an interest his family fortune allowed him to indulge. He filled his home with all manner of electrical experiments, including more than a mile of copper wire strung between the trees on his estate to capture the power of lightning. His superstitious neighbors, seeing the lightning crackle around the wires and hearing the sharp snap and bang of electric batteries discharging, suspected he was completely mad.

Among his experiments was an attempt to unite the sciences of geology and galvanism by using electrical current to induce the growth of quartz crystals. In his music room he fashioned a device that continuously dripped an acidic solution over an electrified stone. Crosse hoped crystals would form on the stone, but this never happened. What happened instead was much stranger. His own words tell the story well:

> On the fourteenth day from the commencement of this experiment I observed through a lens a few small whitish excrescences or nipples, projecting from about the middle of the electrified stone. On the eighteenth day these projections enlarged, and struck out seven or eight filaments, each of them longer than the hemisphere on which they grew. On the twenty-sixth day these appearances assumed the form of a perfect insect, standing erect on a few bristles which formed its tail. Till this period I had no notion that these appearances were other than an incipient mineral formation. On the twenty-eighth day these little creatures moved their legs. I must now say that I was not a little astonished. After a few days they detached themselves from the stone, and moved about at pleasure.

For weeks Crosse watched perplexed as insects multiplied and squirmed around his experiment until they numbered in the hundreds. He repeated the experiment and got the same result—more insects. But being the respectable Englishman that he was, he didn't want to leap to conclusions. Specifically, he hesitated to claim that his experiment had somehow brought forth a new form of life. But a visiting publisher got wind of what had happened and claimed this for him, announcing the news in the local paper under the headline EXTRAORDINARY EXPERIMENT. The media dubbed the insects *Acarus crossii*, in his honor.

Once word of the experiment got out, Crosse's neighbors decided he was not only mad, but quite possibly a devil worshipper as well. In the ensuing months he received numerous death threats. He was called a Frankenstein and a "reviler of our holy religion." Local farmers claimed his insects had escaped and were ravaging their crops, and a priest performed an exorcism on the hill above his house.

Ironically, though his electrical-insect experiments occurred long after the publication of Shelley's novel, it is possible that Crosse was the original role model for the character of Victor Frankenstein. Twenty-two years earlier, in 1814, he had delivered a lecture in London on "Electricity and the Elements." He described the network of wires strung around his country estate that allowed him to conduct bolts of lightning into his house. Sitting in the audience was a young Mary Shelley. His speech reportedly made a great impression on her.

Meanwhile, in 1836, the British scientific community didn't know what to make of Crosse's discovery. A few, such as Cambridge geology professor Adam Sedgwick, angrily denounced it. But others were intrigued. The surgeon William Henry Weekes repeated the experiment, and after a year claimed to have obtained "five perfect insects." But four other researchers—John George Children, Golding Bird, Henry Noad, and Alfred Smee—repeated it and obtained nothing. Likewise, the esteemed biologist Richard Owen examined the insects and pronounced them nothing more than common cheese mites. That judgment pretty much ended the debate over the electrical *Acarus crossii*. They were downgraded from an extraordinary discovery to a common pest.

More than one hundred years later, in 1953, two researchers at the University of Chicago performed an experiment in a similar vein. Stanley Miller and Harold Urey combined water, methane, ammonia, and hydrogen in a flask, and subjected this chemical brew to periodic electrical discharges.

Their goal was to mimic the atmospheric conditions thought to exist on the primitive earth, to see whether the building blocks of life would emerge. They did. Within a week Miller and Urey found high concentrations of organic compounds, including many of the amino acids that form proteins in living cells. However, they reported no sign of cheese mites. Andrew Crosse would have been disappointed.

Crosse, A. (1841). "Description of some experiments made with the voltaic battery . . . for the purpose of producing crystals; in the process of which experiments certain insects constantly appeared." *Transactions and Proceedings of the London Electrical Society* 1: 10–16.

Severed Heads—an Abbreviated History

The weighted blade of the guillotine crashes down and with a whack severs the neck. Another head rolls into the executioner's basket.

The French Revolution and the decades following it were productive years for the guillotine. But as the heads piled higher, a disturbing question formed in the minds of onlookers. Did those decapitated heads retain consciousness for any length of time? Were they aware of what had happened to them? Amateur researchers tried yelling at the heads to see whether they could get a response, but such efforts proved futile. However, they did inspire men of science to ponder a more far-reaching question: Could a head be made to survive isolated from the body? Having thought of the question, they were determined to find an answer.

In 1812 the French physiologist Julian Jean Cesar Legallois

speculated that an isolated head might survive if provided with a supply of blood, but it wasn't until 1857 that his theory was put to the test. Dr. Charles Édouard Brown-Séquard lopped off the head of a dog, drained its blood, and after ten minutes injected fresh blood back into the arteries. Soon, he reported, the severed head stirred to life, displaying what appeared to be voluntary movements in the eyes and face. This continued for a few minutes until the head once again died, accompanied by "tremors of anguish."

Isolated-head research continued with Dr. Jean-Baptiste Vincent Laborde, a man whose brain weighed exactly 1,234 grams. We know this because Laborde was a member of the colorfully named Society of Mutual Autopsy. This society was a social club with one purpose—dissecting one another's brains. Thankfully, the group waited until a member died of natural causes to perform the dissection. Laborde's brain caused a bit of gossip because it turned out to be somewhat light. (The average brain weighs approximately three pounds, or 1,360 grams.) Had he just been posing as an intellectual heavyweight all those years? His friends, eager to preserve his reputation, insisted his brain must have shriveled because of old age.

In 1884, long before his gray matter was removed and weighed, Laborde became the first scientist to perfuse a severed human head with blood. The head belonged to a murderer named Campi (nineteenth-century newspapers tended to refer to all criminals by single names, like modern-day pop stars) and came courtesy of the French authorities. The results were disappointing—nothing much happened, a fact Laborde blamed on the hour-long delay between Campi's execution and the delivery of his head to the lab. But according to rumor, Campi's skin was later removed and used to blind the copies of his postmortem examination. So the experiment wasn't a total loss.

Laborde subsequently conducted a more successful trial on the murderer Gagny, whose head he received only seven minutes after execution. By the eighteen-minute mark he connected Gagny's carotid artery to the corresponding artery of a still-living dog, and blood was pumping through it. Laborde reported that the facial muscles contracted, as though the man were still alive, while the jaw snapped violently shut. But unfortunately (or fortunately for Gagny), no signs of consciousness appeared.

Around the same time, one of Laborde's colleagues, Paul Loye, attempted to settle the debate about postguillotine consciousness by erecting a guillotine in the offices of the Sorbonne and using it to decapitate hundreds of dogs. He assembled a second-by-second chronology of the canine response to sudden head loss, a subject surely never again to be studied as thoroughly. He concluded that the guillotine caused almost instantaneous loss of consciousness, although signs of facial agitation, including dilation of the nostrils and opening and closing of the mouth in what resembled a yawn, persisted for up to two minutes.

After Laborde, a handful of doctors pursued similar research, but for a real breakthrough in severed-head studies the world had to wait until the late 1920s. That's when Soviet physician Sergei Brukhonenko succeeded in keeping the isolated head of a dog alive for over three hours. What made this possible was the use of anticoagulant drugs and a primitive heart-lung machine developed by Brukhonenko. He called it an autojector.

Brukhonenko displayed one of his living dog heads in 1928 before an international audience of scientists at the Third Congress of Physiologists of the USSR. As part of the demonstration, he showed that the severed head reacted to a variety of stimuli. It flinched at loud noises such as a hammer banging on the table beside it. The pupils contracted when light was shone in them. It licked citric acid off its lips. And it even swallowed a piece of cheese, which promptly popped out the esophageal tube on the other end.

Brukhonenko's severed dog heads became the talk of Europe. The playwright George Bernard Shaw wrote a letter to the *Berliner Tageblatt* suggesting, apparently quite seriously, that Brukhonenko's technique be used to extend the life of scientists suffering from terminal disease. He mused, "I am even tempted to have my own head cut off so that I can continue to dictate plays and books without being bothered by illness, without having to dress and undress, without having to eat, without having anything else to do other than to produce masterpieces of dramatic art and literature." He also imagined doctors removing professors' failing bodies and allowing their brains to live on as pure intellect. An entire university, he proposed, could be chaired by bodiless heads.

Shaw's idea is an intriguing one. Faculty housing certainly wouldn't be a problem at such an institution. And it would give new meaning to "going to the head of the class." However, many people might understandably hesitate before volunteering to become a part of the student body.

Brukhonenko, S. S., & S. Tchetchuline (1929). Expériences avec la tête isolée du chien. *Journal de physiologie et de pathologie générale* 27 (1): 31–45.

Human-Ape Hybrid

Dr. Il'ya Ivanov was frustrated. He believed his research was of great, possibly world-shaking, significance. If successful it would make him one of the most famous men in the world. And yet here he was, thousands of miles from European civilization, reduced to sneaking around a West African research station like a criminal, hiding his intentions from the suspicious native staff. Only his son knew his true purpose. Together they planned to create a new kind of creature—a human-ape hybrid.

Early in the morning of February 28, 1927, the father-and-son team told the staff they would be inspecting two female chimps, Babette and Syvette, for medical treatment. They knew they didn't have a lot of time. If the staff realized what they were actually doing, Ivanov wrote in his notebook, he and his son would face "very unpleasant consequences." So, much to his displeasure, the insemination would have to be done fast. His son carried a gun in his pocket, in case the chimps fought back.

Ivanov and his son subdued the chimps and prepared to place human sperm inside the uteri of the animals. They used the tools of artificial insemination developed by the elder Ivanov in Russia, where his years of research had revolutionized the field of veterinary reproductive biology and had set the stage for the rise of large-scale stud farming there. However, the procedure went badly. Feeling rushed, Ivanov failed to fully insert the sperm. He knew there was little chance of success.

For many decades Ivanov's gruesome hybridization experiments remained little known in the West. There were rumors, but few concrete details. Ivanov never published his findings. It was only after the collapse of the Soviet Union and the opening of Russian archives that details finally emerged.

The militantly secular Soviet government sponsored Ivanov's research, believing that a successful human-ape hybrid would have, if nothing else, enormous symbolic significance. This was less than two years after the Scopes Monkey Trial had demonstrated the hostility with which many Christian fundamentalists in the United States greeted any suggestion of an evolutionary relationship between man and apes. The pro-Darwin, Marxist leaders of the Soviet Union rubbed their hands together with glee at the thought of presenting the fundamentalists with a "human-zee."

But Ivanov received aid from other sources as well. The French Institut Pasteur, fully aware of his plan, provided him with access to their West Guinea research facility, hopeful that his work would contribute to scientific understanding of the origins of man.

Later in 1927, Ivanov made one more attempt to impregnate a female chimp with human sperm, but this third try proved no more successful than the first two. He knew from his work with livestock that five or six inseminations per animal provided the optimal chance of success, but social conditions at the research facility didn't allow him that luxury. None of the chimps ever showed signs of pregnancy.

Faced with failure, Ivanov turned to Plan B—impregnate human women with ape sperm. He made inquiries at a Congo hospital about the possibility of inseminating female patients. He suggested it would be prudent to do so without the women's knowledge. His request was denied. Disheartened, and complaining about Africa's "backward" culture, he returned to the Soviet Union, where he hoped to continue his experiments.

He brought back a male orangutan named Tarzan to serve as the sperm donor. He also revised his plan, deciding to seek out female volunteers. Remarkably, he got a few. One woman cheerily wrote to him that she was willing to surrender her body to science because "I don't see any sense in my further existence." Once again, though, fortune did not favor Ivanov. Tarzan died of a brain hemorrhage in 1929 before the experiment could start, leaving Ivanov apeless. The next year Ivanov was swept up in one of Stalin's political purges and shipped off to a prison camp. He was released two years later, but died soon thereafter. This, as far as we know, brought an end to his research program.

Ivanov's experiments mark a low point in the history of biological research. But they raise an interesting question. Could he have succeeded? Is a human-ape hybrid possible?

Humans are very closely related to other primate species, chimpanzees in particular. We share 99.4 percent of our DNA with them. The phrase "human-ape hybrid" is itself misleading, since humans are, in fact, a species of African ape. A May 2006 study published in *Nature* speculated that after humans split from chimps between five and seven million years ago, human evolution may have been influenced by continued interbreeding with chimpanzees. Many biologists see no reason why a human-chimp match would not still be possible, though the topic remains controversial.

And just in case you're curious: No, Ivanov did not use his own sperm during the 1927 experiments. The identity of that proud father-to-be remains unknown. Ivanov only identified him as "a man whose age isn't exactly known. At least, not older than thirty."

Rossiianov, K. (2002). "Beyond species: Il'ya Ivanov and his experiments on cross-breeding humans with anthropoid apes." *Science in Context* 15 (2): 277–316.

The Man Who Cheated Death

"I will have the formula that will start the blood circulating again, and with it breath, and with it life!"

So spoke Dr. John Kendrick, a character in *Life Returns*, a 1930s B movie. Kendrick was fictional, but he was based on an actual person—Berkeley scientist Robert E. Cornish, a man who achieved notoriety by claiming he could defeat death.

Cornish's career got off to a promising start. He was a child prodigy, graduating from the University of California with honors at the age of eighteen and receiving a doctorate by the time he was twenty-two. He then accepted a position at the University of California's Institute of Experimental Biology, where he worked on projects such as lenses that made it possible to read a newspaper underwater. For some reason, they never caught on. But in 1932, while still only twenty-seven years old, he became obsessed by the idea that he could restore life to the dead.

At the heart of Cornish's plan was a teeter board. This was essentially a seesaw. "By tying the 'dead' subject to a teeter board, and alternately tipping up and down," Cornish wrote, "one expects a considerable artificial circulation of the blood." His theory was that if you could get the blood flowing in recently deceased patients who had suffered no major organ damage, life would return.

During 1933 he attempted to revive victims of heart attack, drowning, and electrocution with the teeter board, but had no success. He did note, in a confidential report submitted to the University of California, that after the corpse of a heart-attack victim was "teetered" for over an hour, the "face seemed to warm up suddenly, sparkle returned to eyes, and pulsations were observed in soft tissue between windpipe." But the guy remained dead.

Cornish decided to perfect his method on animals before trying it again on humans. In 1934 he went public with a series of canine resuscitation experiments. He operated on a total of four fox terriers, naming them, in an allusion to the biblical character brought back to life by Jesus, Lazarus II, III, IV, and V. The fate of Lazarus I was not recorded.

First he killed the terriers, asphyxiating them with a mixture of nitrogen and ether until their heartbeats and breathing stopped. Then he tried to revive them using a combination of teetering, mouth-to-snout resuscitation, and injections of adrenaline and heparin (an anticoagulant).

Amazingly, he had some success. The dogs returned to life. The catch was that it was a meager semblance of life. Lazarus II and III died (again) after a few hours, having never achieved consciousness. Lazarus IV and V were more of a success. They lived on for months, though blind and severely brain damaged. It was said they inspired terror in other dogs they met.

The press ate up the news of Cornish's research, delivering blow-by-blow accounts of each experiment. "I could hear the breath coming back into that still body," one reporter wrote of Lazarus II. "Slowly at first, then quickly as if the dog were running. The legs twitched. Later I heard a whine and a feeble bark." It helped that with his brooding eyes, sallow skin, and dark hair, Cornish looked the part of a mad scientist.

Hollywood also loved Cornish. Universal produced *Life Returns* (quoted from above) in 1935. It would be a totally forgettable movie—imagine a bad combination of *Frankenstein* and *Our Gang*—except that it features five minutes of Cornish's actual experiments spliced into the action. Cornish's work also inspired a number of Boris Karloff movies, including *The Man with Nine Lives* and *The Man They Could Not Hang*.

The University of California, however, was not so taken with Cornish's new line of research. Faced with complaints from animal-rights activists, the school ordered him off its campus and severed all ties with him. He retreated to his Berkeley home.

Cornish lay low for the next thirteen years, fending off hostile neighbors who complained about sheep and dogs escaping from his lab and mystery fumes that made paint peel on surrounding buildings. But in 1947 he triumphantly returned to the headlines with news that he had perfected his technique and was ready for a bold new experiment. He would bring an executed prisoner back to life! He had moved on from teeter boards. Now he unveiled a Heath Robinson–style heart-lung machine made out of a vacuum cleaner blower, radiator tubing, an iron wheel, rollers, and a glass tube filled with sixty thousand shoelace eyes.

San Quentin death-row prisoner Thomas McMonigle volunteered to be Cornish's guinea pig—despite assurances that, if the experiment was successful, he would still have to remain in prison—but the experiment was never given a chance. California state authorities flatly turned down Cornish's request.

Utterly defeated, Cornish returned home and eked out a living selling a toothpaste of his own invention, Dr. Cornish's Tooth Powder. He died of a stroke in 1963. The local paper noted in his obituary that while attending Berkeley High School as a teenager he had been the "first student ever known to wear sandals to school regularly." It was a fitting tribute to a man who never quite fit in.

Cornish, R. E., & H. J. Henriques (1933). "Report of Investigation of Resuscitation." Unpublished manuscript.

The Two-Headed Dogs of Dr. Demikhov

A hiker wandering through the forests outside of Moscow comes across a large, official-looking building. Peering over the fence surrounding it, he sees doctors and nurses walking dogs around a courtyard. Hardly a shocking sight. But a second look leaves the hiker puzzled, and scared. There's something different about these animals. He sees a dog limp by with one leg a conspicuously different color than the rest of its body—as though the leg had been sewn on. And could it be? Surely not. But yes! One of the other dogs has two heads.

The Soviet Union shocked the world in 1954 when its government proudly unveiled a two-headed dog. The strange animal was the creation of Vladimir Demikhov, one of the nation's top surgeons. He had honed his craft in field hospitals during World War II, after which the government set him up in a top-secret research center outside Moscow. His mission there was to prove the Soviet Union's surgical preeminence.

Demikhov created the two-headed dog by grafting the

head, shoulders, and front legs of a puppy onto the neck of a mature German shepherd. Eventually he created twenty of these hybrids. However, because of postoperative infection, most of the dogs didn't live long. The record was twenty-nine days—suggesting that, at least as far as the dogs were concerned, two heads were not better than one.

The dogs made headlines around the world. The press nicknamed them Russia's "surgical Sputnik." In 1959 United Press reporter Aline Mosby visited Demikhov's lab and met Pirat, a German shepherd/puppy combo. Accompanying Demikhov on a walk with Pirat, she noted Pirat had to be led by the ears because a normal collar wouldn't fit around his neck.

Mosby also reported that although the two heads shared a circulatory system, they led separate lives. They slept and woke at different times. The puppy even ate and drank on its own, though it didn't need to because it received all its nourishment from Pirat. When the puppy eagerly lapped at a bowl of milk, whatever went into its mouth dribbled out the stump of its esophageal tube onto Pirat's shaved neck.

Was there any medical justification for the dogs? Critics didn't think so. They dismissed them as a publicity stunt. Demikhov, however, argued that they were part of a continuing series of experiments in surgical techniques. His ultimate goal was to make possible a human heart-and-lung transplant. In fact, another doctor eventually performed the first human heart transplant—Dr. Christiaan Barnard in 1967—but Demikhov is widely credited with paving the way for it.

Demikhov also envisioned a future in which banks of surgical spare parts could be created by grafting extra sets of limbs onto human "vegetables"—his term for brain-dead patients. When needed, the limbs would be removed. An entire market in used extremities could come into existence. However, Demikhov seriously underestimated the problems involved with tissue rejection. For that reason, you don't need to fear a Demikhov Limb and Organ Bank opening on a street corner near you anytime soon.

Demikhov, V. P. (1962). *Experimental Transplantation of Vital Organs*. New York: Consultants Bureau.

Franken-Monkey

The monkey opened his eyes. Even through the haze of drugs, he could sense something was wrong. He tried to move but couldn't. Why were his limbs not responding? He felt scared and wanted to run. Instead he could only stare straight ahead. What was this place he was in? Who were these men that surrounded him? Angrily he tracked their movements with his eyes and warned them away the only way he was able—by baring his teeth and snapping menacingly at the empty air.

When American leaders learned that Vladimir Demikhov had created a two-headed dog, they knew they had to respond. For the sake of national pride, they not only had to match Demikhov's achievement, but also had to do one better. Thus ensued one of the more peculiar chapters of the Cold War— a surgical arms race. Though perhaps *head race* would be a more fitting term.

America's answer to Demikhov was Robert White. In 1960 White was a thirty-four-year-old Harvard-trained surgeon with great ambitions. He wanted to make a name for himself, and if in doing so he could simultaneously help his country, then all the better. So in 1961, with the help of the

U.S. government, he established a brain research center in Cleveland, Ohio. The government told him to do whatever it took to beat Demikhov.

White agreed with critics who thought Demikhov's dogs were a bit of a stunt. Sensational, yes. But still a stunt. After all, stitching the upper body of a puppy onto the neck of an adult dog was not a true head transplant. What White envisioned doing was altogether more ambitious. He would cut the head off an animal and then sew on a new, functioning head. It would be a true head transplant, the kind of thing found only in Hollywood movies and science-fiction novels.

But before he could do this, he had to learn more about how the brain functioned. This would take him years of study and experimentation.

Step one in this process was to find out whether a brain could be isolated from the body and remain alive. On January 17, 1962, he proved this could be done. He removed the brain of a monkey from its skull and sat it on a stand, supplied with blood from an external source. This was a far more complicated procedure than simply lopping off the top of the skull and lifting out the gray matter, because the arteries supplying blood to the brain had to remain intact. White had to carve away the tissue of the face—the skin, nerves, muscle, and cartilage—until all that remained was the skull attached to the body by the thread of the arteries. Only then did he crack open the skull and reveal the brain. It took hours. As he worked, he puffed on a pipe and chatted about current affairs, as though he were chiseling away at a piece of wood instead of a living creature.

The brain sat motionless on the stand, a gray mass of tissue. Only by its electrical activity—the blips of an EEG trace—could one tell it was alive and thinking. After a couple of hours, having done what he set out to do, White switched off its blood supply. It took three minutes for the brain to die.

The next step was to find out whether a brain could survive being transplanted into another living creature. White achieved this goal on June 3, 1964. He removed the brain of a dog and placed it under the neck skin of another dog, where the brain remained alive, floating in darkness, for days. Unfortunately for the dog that played host, it was no smarter for having a second brain. In fact, the extra brain was literally nothing more than a pain in the neck.

The final step in White's research program was a full head transplant. Six more years of preparation were necessary, but on March 14, 1970, White did it. In a carefully choreographed operation requiring a large team of assistants, he separated a monkey's head from its body and reattached the head to a new body. After a few hours the monkey woke up to its new reality. White wrote that it "gave evidence of its awareness of the external environment by accepting and attempting to chew or swallow food placed in its mouth. The eyes tracked the movement of individuals and objects brought into their visual fields." When White placed his finger in the monkey's mouth, the monkey bit it. Evidently, it wasn't a happy monkey.

It's hard to imagine an experience more disorienting than waking up and discovering you have a new body, but it could have been worse for the monkey. White could have placed the head on the body the wrong way around. He noted that, because of the way the two bodies were positioned in the lab, it would have been far easier for him to do this, but for the monkey's sake, he didn't. As if it really mattered to the monkey at that point.

The monkey couldn't get up and walk around or swing from trees. Although the head was attached to a new body, it couldn't control that body in any way. The spinal cord remained severed. The monkey was now a quadriplegic. In essence, the new body was merely a pump supplying blood to the head. From a surgical point of view, it was an impressive piece of work. But it seems a mercy that the monkey survived only a day and a half before succumbing to complications from the surgery.

White had achieved his goal, but at a personal cost. Instead of hailing him as a national hero, the public was appalled by his work. Funding for his experiments gradually dried up. But White was hardly one to back down. Instead, he played his role of a modern-day Dr. Frankenstein to the hilt. He freely admitted during interviews that he was a fan of the Frankenstein movies. He once showed up on a children's TV program toting a Dr. Frankenstein's doctor bag. He even publicly lobbied for the need to take his work to the next level—a human head transplant. He argued that if doctors were willing to replace a patient's heart, why not replace the entire body? Surgically, it was possible. And if the patient was already a quadriplegic, it wouldn't significantly alter his lifestyle. He toured with Craig Vetovitz, a near-quadriplegic who volunteered to be his first head-transplant patient.

White admitted there was a long way to go before the public was ready to accept the idea of full-body transplants, but he predicted that "the Frankenstein legend, in which an entire human being is constructed by sewing various body parts together, will become a clinical reality early in the twenty-first century." If he's right, eventually the public will have to get its head around the idea.

———

White, R. J., et al. (1971). "Cephalic exchange transplantation in the monkey." *Surgery* 70 (1): 135–39.

CHAPTER TWO

Sensorama

Morton Heilig's Sensorama, built in 1957, was the first fully immersive virtual reality machine. Users sat on a vibrating seat as they viewed 3-D movies. Fans blew wind through their hair; speakers played simulated road sounds; and canisters sprayed the scents of fresh-cut grass and flowers into the air around them. All of this created the illusion users were riding through the countryside on a motorcycle. The Sensorama gives its name to this chapter because we now embark on a journey through the peculiar and often unnerving world of sensory research.

We will examine experiments that probe the mysteries of touch, taste, smell, sight, and sound. Like the Sensorama, a few of these experiments had a commercial motive. Most of them, however, were inspired by a deeper philosophical principle that can be summed up by the thirteenth-century philosopher Thomas Aquinas's peripatetic axiom: "Nihil est in intellectu quod non prius in sensu." In English: There is nothing in the mind that is not first in the senses. We gain knowledge through sense-based experience. Therefore, to understand human knowledge, we must first understand the senses and how they distort or enhance the world around us. As we will see, the emphasis seems to be on distortion rather than enhancement.

1. TOUCH

The Mock-Tickle Machine

A blindfolded man sits in a chair. His bare foot, strapped to a stool, rests inches away from a robotic hand connected to an array of rubber hoses and controls. A woman in a lab coat sits down next to him. "You will be tickled twice," she states without emotion. "First I will tickle you, and then the tickle machine will have its turn." As she says this she glances at the robotic hand. The man nods his understanding.

No, this isn't a scene from a fetish club. The setting is the UC San Diego psychology lab of Dr. Christine Harris. During the late 1990s thirty-five undergraduates agreed to bare their feet for Dr. Harris and endure tickle-torture to help her answer that age-old question, Why can't we tickle ourselves?

Two contradictory answers to this question had previously been proposed. Theory one (the interpersonal theory): Tickling is a social act and requires the touch of another person to elicit a response. Theory two (the reflex theory): Tickling is a reflex that depends on unpredictability and surprise. We can't tickle ourselves, proponents of this theory argue, because we can't surprise ourselves.

Harris designed her tickle-machine experiment to put these opposing theories to the test. If the interpersonal theory was correct, she reasoned, a machine-tickle should not be able to elicit laughter from a person. But if a machine could generate this response, that would imply the reflex theory was correct.

So the students came to her lab, took off their shoes and socks, and waited to be tickled. As they sat blindfolded, they felt themselves being tickled by the experimenter (Harris) and then by the machine. Their response to each stimulus was rated on a scale of zero (not at all ticklish) to seven (very ticklish).

However, not all was as it appeared. Unbeknownst to the students, neither Harris nor the machine tickled anyone. The tickle machine wasn't even capable of tickling anyone. It was just a stage prop that made a loud vibrating sound when turned on. The students were actually being tickled by a woman hiding beneath one of the cloth-covered tables in the lab.

Upon receiving the signal, the secret tickler—aka "research assistant"—lifted the tablecloth, reached out her hand, surreptitiously tickled the foot of the subject, and then retreated back into her lair.

Once the assistant caught her hair in the top of the table, and as she struggled to free herself, the student test-subject realized that something strange was going on. Otherwise, the deception worked perfectly. The tickler did her job so well that during follow-up questioning a number of subjects commented on the artificial feel of the tickle machine. One student said the "machine felt unnatural, temperature different. Like walking across plush carpet."

Why the elaborate ruse? Harris could have created a real,

working tickle machine. A British researcher, Sarah Blakemore, later did exactly this—although Blakemore was looking at the kind of tickle that is an itchy feeling, like a bug crawling across your skin, whereas Harris was investigating the intense, laughter-eliciting kind of tickle. Harris was concerned that a robotic tickler might feel different than a human one, and she didn't want this difference in tactile quality to influence her results. All she needed was for participants to believe a machine was tickling them. If they truly believed this, then, according to the interpersonal theory, they should not respond to the tickling.

Harris found the students laughed just as hard when they believed they were being tickled by a machine as when they believed a human to be the source of the sensation. This led her to conclude that the reflex theory was correct. In her words, "The tickle response is some form of innate stereotyped motor behavior, perhaps akin to a reflex." Which is a useful fact to know the next time the subject comes up at a cocktail party.

You might also be relieved to learn that the research assistant has finally been allowed out from under the table. Though if you ever visit the Harris lab you might want to take a quick peek beneath the tables, just to be on the safe side.

_____

Harris, C., & N. Christenfeld (1999). "Can a machine tickle?" *Psychonomic Bulletin & Review* 6 (3): 504–10.

Touching Strangers

You reach out and touch a stranger. As your fingers brush against his arm, it's like an electric current passing between the two of you. You become aware of the texture of his skin, the warmth of his body, and his proximity to you. But you also feel an undercurrent of apprehension. How will he react to your touch? Will he interpret the gesture as friendly? Erotic? Comforting? Condescending? What about threatening?

Colin Silverthorne, a professor of psychology at the University of San Francisco, discovered the power and danger of touching a stranger during an experiment he conducted in 1972. He told subjects the purpose of his research was to gather information on the aesthetic value of a series of pictures. But the real purpose was to see whether being touched by the researcher would translate into a greater appreciation of whatever they were viewing.

At one point during the presentation, while pretending to adjust the focus of the slide projector, Silverthorne casually placed his hand on each subject's shoulder for three seconds. Most people didn't seem to notice the touch, but they did give higher marks to the pictures seen at the moment of contact, except for one woman. When Silverthorne's hand landed on her shoulder, she "showed an agitated response" so extreme that she had to be excused from the experiment. Evidently, some people don't like strangers touching them.

However, there is one situation in which the touch of a stranger is almost always well received—in restaurants. In 1984 researchers April Crusco and Christopher Wetzel enlisted the help of waitresses at two restaurants in Oxford, Mississippi, to investigate the effect of touch on tipping. When returning with a customer's change, a waitress performed one of three "touch manipulations." She either briefly touched the diner's palm while delivering the change, placed a hand on the customer's shoulder, or didn't touch the person at all. The waitresses did not vary any other aspect of their behavior:

> The waitresses approached the customers from their sides or from slightly behind them, made contact but did not smile as they spoke "Here's your change" in a friendly but firm tone, bent their bodies at an approximately 10 degree angle as they returned the change, and did not make eye contact during the touch manipulation.

The researchers collected data on 114 diners, most of them college students, none of whom suspected they were the subject of an experiment. The no-touch condition generated the smallest tips. A touch on the shoulder earned 18 percent more, and a fleeting touch on the palm garnered a full 37 percent bonus. Clearly, it paid to touch the diners.

Follow-up experiments by other researchers have confirmed these results, while simultaneously exploring other variables that may play a role in the touch-tipping effect. For instance, a 1986 experiment in Greensboro, North Carolina, focused on the element of gender. When serving a man and a woman dining together, would a waitress earn more by touching the man or the woman? The answer was the woman. The investigators theorized that a female server touching the male diner generated jealousy from his companion. However, touching the man still led to a higher tip than touching neither diner.

A 1992 Chicago experiment included the attractiveness of the server as a variable. The largest tip earners turned out to be highly attractive waitresses who touched female diners. They received a full 41 percent more, on average, than the lowest earners, unattractive men who touched no one. The researcher measured attractiveness by prior customer surveys and didn't inform the servers that their good looks (or lack thereof) played a part in the study. Breaking the bad news to the waiters rated ugliest-of-the-bunch doubtless would have been a bit cruel.

Tip-maximization strategies turn out to be a popular subject for research—perhaps a case of scientists preparing for hard times in case their next grant application doesn't get funded. Thanks to the tireless efforts of experimenters, we know that servers can increase their tips by, in addition to touching customers, introducing themselves, being friendly, kneeling down during interactions, smiling a lot, immediately repeating a diner's order, wearing flowers in their hair (if female), giving a joke-bearing card or a gift of candy with the check (the more candy, the better), writing a patriotic message such as "United We Stand" or a smiley face on the check, and, most important of all, showing up for work when either the moon is full or the sun is shining. A 1996 experiment even found that if a waitress works in an Atlantic City casino where the outside weather conditions can't be seen, it behooves her to tell patrons it's sunny outside, though it might be pouring rain.

Waiters may also want to remember a tip-maximization strategy that has received less attention from researchers—providing prompt, attentive service.

Crusco, A. H., & C. G. Wetzel (1984). "The Midas Touch: The Effects of Interpersonal Touch on Restaurant Tipping." *Personality and Social Psychology Bulletin* 10 (4): 512–17.

2. TASTE

What's the Difference?

The wine connoisseur picks up the glass of red wine and holds it to the light. "Deep mahogany," he mutters. He swirls the glass, noting how the liquid clings to the sides. Then he sniffs it, deeply inhaling the bouquet. "A touch of ground coffee, spices, leather, and black currants," he murmurs. Finally, he sips the wine. He swirls it around in his mouth and allows it to linger on his tongue. He savors the experience, and only then does he let the liquid slide down his throat.

He places the wineglass on the table. "A full, rounded body," he proclaims. "Good fruit. Cloves and sweet toffee flavors. Lovely finish. An excellent vintage. My guess—Châteauneuf, 1989."

Wine connoisseurs can put on quite a show. But is there anything to it? Can they really pinpoint the exact vintage of a wine simply by tasting it? If they do make this claim, they'd better hope they don't find themselves in one of Frédéric Brochet's experiments, because Brochet has a way of making wine connoisseurs look like fools.

Brochet was a cognitive neuroscience researcher at the University of Bordeaux. In 1998 he invited fifty-four specialists to taste some wines and write down their impressions. First he served a red and a white. The tasters scribbled down their notes. Next he served a different red and white. Again, they jotted down comments. To describe the two reds they used terms such as *plump, deep, dark, black currant, cherry, fruit, raspberry,* and *spice.* The two white wines evoked descriptors such as *golden, floral, pale, dry, apricot, lemon, honey, straw,* and *lively.* Both sets of adjectives are commonly used in the wine industry, with specific reference to either reds or whites.

The specialists thought they were just tasting some wine. Little did they know they were the subjects in an experiment to find out whether connoisseurs can tell the difference between red and white wine.

Unbeknownst to all the specialists, the second set of wines they tasted, the red and the white, were identical. Brochet had simply added flavorless food coloring to some of the white wine to create a faux red. One would think that if connoisseurs' palettes are sensitive enough to allow them to detect the exact vintage of a wine, then they should have no problem telling when they've been served two glasses of the same wine—even if one of the glasses has a bit of red dye in it. But no. Not a single person wrote down that the second pair of wines tasted similar, nor that the "red" tasted like a white. Their descriptions of the dyed white read exactly like descriptions of a red wine. The inescapable conclusion was that the specialists had all been fooled.

A follow-up experiment proved just as bruising to the egos of wine connoisseurs. Brochet told his group of specialists he would serve them two wines, the first a common table wine and the second a premium vintage.

He showed them the table wine, poured a sample into their glasses, and took a sip himself. He promptly spat it out dismissively. The tasters then tried it and wrote down their impressions—*simple, unbalanced, light, fluid,* and *volatile.*

Next Brochet showed them the premium wine. He took a long sip and smacked his lips in appreciation. When it was their turn, the tasters described the wine as *complex, balanced, flavorsome, smoky, fresh, woody,* and *excellent.* You can probably guess the punchline. Again, the two wines were identical. They were both the same common Bordeaux.

Does this prove that wine connoisseurs are full of hot air? Are they unable to tell the difference between the good stuff and plonk, or even between red and white? Not quite, though it would be easy to interpret Brochet's experiments in that way.

Brochet didn't design his studies to knock wine connoisseurs down a peg. He himself is a wine lover and founder of the Ampelidae winery in the west of France. He argues that his experiments instead demonstrate the power of perceptive expectation: "The subject perceives, in reality, what he or she has pre-perceived and finds it difficult to back away."

What this means is that the brain does not treat taste as a discrete sensation. Instead, it constructs the experience of flavor by taking into consideration information from all the senses—sight, sound, smell, touch, and taste. Paradoxically, it places the greatest emphasis on sight—almost twenty times more emphasis, according to Brochet, than it places on any other sense. So if our eyes tell us there's red wine in the glass, our brain places more faith in that data than in the information coming from the taste buds. Our expectation becomes our reality.

Ironically, the more highly trained a wine drinker is, the more likely it is he'll fall for the red-dye-in-white-wine trick. This is because connoisseurs are highly conditioned to expect a red-colored wine to taste a certain way. They can't escape their preconceptions.

Does this mean you can pour some swill into fancy bottles, serve it at your next party, and no one will be the wiser? Sure. Give it a try. But ask yourself this: How do you know the same trick hasn't already been played on you? Brochet points out that almost all wine fraud is exposed by faulty paperwork, not by consumers complaining about the taste. In other words, just how good *was* that cabernet you paid fifty dollars for last month?

Brochet, F. (2001). "Chemical object representation in the field of consciousness." Académie Amorim. Unpublished manuscript. Available: http://www.academie-amorim.com/

Coke vs. Pepsi

It's one thing to suggest wine connoisseurs may not be able to tell the difference between red and white wine. But try telling Coke and Pepsi fans that their beloved soft drinks are indistinguishable. You'll soon have an angry mob on your hands, begging to differ.

The foolhardy researcher who dared question the cola dogma was Read Montague of Baylor College of Medicine. In 2005 Montague conducted a scientifically controlled, double-blind version of the Pepsi challenge. Participants received two unlabeled cups containing Coke and Pepsi. They were asked to drink them and indicate which tasted better. The result—an even split between the two drinks, with no correlation between the brand of cola participants claimed to prefer beforehand and the one they chose in the study. Tasters could not distinguish between the two. These results horrify Coke and Pepsi lovers. They insist—science and double-blind tests be damned—that *they* would have been able to tell the difference.

But it gets worse for cola fans, because Montague then

took his experiment a step further. He served participants two cups—one labeled Coke, the other unlabeled. Subjects stated a preference for what was in the cup labeled Coke almost 85 percent of the time. The catch was that both cups contained Coke. Apparently Coke with a label tastes better than Coke without one. When Montague next served Pepsi in labeled and unlabeled cups, he found no similar preference effect. This finding suggests that Coke has a better marketing department, which has succeeded in convincing consumers to prefer drinks served as Coke—no matter what the drinks *actually* taste like. Montague put it this way:

> There are visual images and marketing messages that have insinuated themselves into the nervous systems of humans that consume the drinks. It is possible that these cultural messages perturb taste perception.

Finally Montague placed subjects in an MRI scanner and observed their brains as they drank Coke and Pepsi. Serving them the drinks wasn't easy. A vise that prevented movements larger than two millimeters in any direction held their heads in place. Cooled plastic tubes directed the liquid into their mouths.

When Montague flashed an image of a Coke can on a screen over the subjects' heads before squirting the Coke into their mouths, their brains lit up like Christmas trees. But when he flashed a colored light before serving the Coke, or when he served Pepsi (preceded by a picture of a Pepsi can), brain activity was far less. In other words, Coca-Cola's advertising had a measurable effect on neural response.

The slightly creepy implication of Montague's experiment is that advertising can literally rewire the neurons in our head and alter our sensory experience of the world. It can reprogram our perception of reality, compelling us to perceive two near-identical forms of sugary carbonated water as tasting different. So when Coke and Pepsi fans insist there's a difference between the two beverages, for them this is true, because viewing the label has become an inextricable part of the taste experience. This is how their brains are wired. And when others insist there is no difference, that is also true. It's like two species—fizzy-drink lovers and fizzy-drink skeptics— living in incommensurable worlds, constantly arguing past one another. One of those species just happens to suffer from considerably more tooth decay than the other.

Montague, R., et al. (Oct. 14, 2004). "Neural Correlates of Behavioral Preference for Culturally Familiar Drinks." *Neuron* 44: 379–87.

3. SMELL

Synchronous Menstruation (The Scent of a Woman)

It sounds like a bizarre party trick. Take a group of women, allow them to live together for a while—four months should be enough—and then watch as their menstrual cycles, as though guided by an unseen force, begin to synchronize.

Tales used to be told of nuns in convents menstruating in lockstep, but scientists dismissed these stories as folklore not worthy of serious consideration, until the research of a young college student named Martha McClintock forced them to reconsider.

It was the summer of 1968. Following her junior year of college, McClintock was attending a conference at the Jackson Laboratory in Maine. The predominantly male scientists were discussing how female rats housed together ovulated at the same time, a phenomenon attributed to pheromones, airborne chemical signals excreted by the rats. McClintock mentioned that she had seen the same thing happen in humans, in her college dormitory. This comment provoked polite skepticism among the attendees. She was gently told that if she was going to make wild claims, she needed to back them up with proof.

Getting such proof might have seemed like a tall order. After all, it's one thing to study caged ovulating rats, but humans are another matter. McClintock, however, was a student at Wellesley College, a small, all-female liberal arts college outside of Boston. If you wanted the human equivalent of a rats-in-a-cage experimental setting, Wellesley was it.

Next school year, McClintock convinced all 135 of her dorm mates to participate in a study. Three times during the year they recorded data about the date of onset of their recent menstrual cycles, with whom they spent most of their time, and whether they had recently been in the presence of men. A young woman named Hillary Rodham—later better known as Hillary Clinton—happened to be in McClintock's class at Wellesley. Clinton has reported that she lived in Stone-Davis dormitory during her senior year, but McClintock has not publicly revealed in which dormitory she conducted her study.

At the end of the year, McClintock processed the data, and the results were clear. After living together for a year, the women had experienced "a significant increase in synchronization." In other words, at the beginning of the year their cycles were all over the map, but nine months later a high percentage of them were menstruating at almost the same time. The phenomenon was most evident among close friends. This was compelling evidence of the existence of menstrual synchrony in humans.

Subsequent research, such as investigations of lesbian couples and mothers and daughters living together, has largely confirmed McClintock's findings, but her study left unanswered an important question. What causes menstrual synchrony?

McClintock speculated that, just as in rats, the culprit was airborne pheromones—or, put more plainly, *the scent of other women*. But it wasn't until 1998, when she was a professor at the University of Chicago, that she was able to prove it. She instructed nine women to wear sweat-collecting pads in their armpits for eight hours. She then wiped these sweaty pads beneath the noses of twenty other women (who didn't know what it was they were being wiped with). McClintock found that the cycles of the second group altered depending on whether they were wiped with the sweat of a woman late or early in her cycle. She concluded that a chemical messenger in the sweat (i.e., pheromones) was influencing the timing of the other women's menstrual cycles.

This experiment, in turn, raised another question. If pheromones can affect human biology, how do they have this effect? A few scientists have suggested that humans possess a previously unknown sixth sense that detects pheromones. After years of peering up people's nostrils, researchers such as Luis Monti-Bloch and David Berliner declared the discovery of a sixth sense organ—the vomeronasal organ. If they're right, this tiny pit, located about half an inch inside the nose, detects pheromones. It's well known to exist in other animals, but humans were thought to be lacking it. It's so small that most anatomists simply overlooked it.

These discoveries have inspired numerous entrepreneurs to jump on the pheromone bandwagon, advertising pheromone-laced sprays that will, they claim, make their wearer irresistible to the opposite sex. It's doubtful these work. But it's not out of the realm of possibility that one day we might see genuine pheromone-based products on grocery-store shelves. We might even be able to buy "McClintock, the menstrual-cycle altering perfume."

McClintock, M. K. (1971). "Menstrual Synchrony and Suppression." *Nature* 229: 244–45.

The Smell of Money

You've got a drink in your hand. You're popping quarters into the slot-machine. The sound of falling coins rings in your ears. Money being won! Your adrenaline is pumping. And boy, something smells good!

In 1991 Dr. Alan Hirsch introduced two different odors, both rated as pleasant in prior preference studies, into different areas of the gaming floor of the Las Vegas Hilton. The odors were strong enough to be easily perceived, but were not overpowering. A third area he left odor-free. Odorization occurred over a period of forty-eight hours.

The results were startling. One of the odorized areas saw

a 45 percent jump in the amount of money spent at the machines compared to the week before. The second odorized area and the odor-free zone saw no increases. The first odor appeared to have caused people to spend more money—*a lot more money.*

There were no pheromones in these odors. They were simply pleasant aromas. Also, Dr. Hirsch had no idea why the first odor, but not the second, caused the dramatic spike in gaming revenue. He had expected both to have some effect. Nevertheless, the gaming industry and retailers throughout the country immediately took notice. Easy money, they thought, never smelled so good. Pump in a few good scents and wait for the cash to roll in.

Rival researchers, however, criticized Hirsch's work, complaining that he never identified the jackpot scent, making it impossible for them to evaluate his results. Consumer groups, on the other hand, decried the dawning of an era of manipulative smell technology.

Since the early 1990s researchers have continued to investigate the smell-sells phenomenon, but results have been ambiguous at best. Some studies show positive effects, whereas others show none. Marketing professors Paula Fitzgerald Bone and Pam Scholder Ellen, reviewing this research in a 1999 article, cautioned that "evidence is stacked against the proposition that the simple presence of an odor affects a retail customer's behavior." Such words of warning hardly dented the enthusiasm of retailers, who, if anything, have become even more excited about odor in recent years. Some businesses have gone so far as to develop signature scents. Samsung, for instance, fills its stores with a distinctive honeydew melon smell, and Westin hotels pump a white tea fragrance into their lobbies. So the next time you're in a store and you stop to smell the roses—or the melon, vanilla, cucumber, lavender, or citrus—remember that the business owner is hoping soon to be counting your cash.

Hirsch, A. R. (1995). "Effects of Ambient Odors on Slot-Machine Usage in a Las Vegas Casino." *Psychology and Marketing* 12 (7): 585–94.

Smell Illusions

Lift this book to your nose. Its pages have been coated with an odor-producing chemical. Can you smell it? If not, scratch a page to release the scent more fully. The aroma is pleasant and fruity. Can you smell it now? Yes? No?

Well, maybe not. The book contains no scratch-n-sniff odor. At least, it shouldn't. Nevertheless, some readers may, through mere suggestion, have smelled something. Or believed they did.

Suggestion exerts a powerful influence over what we smell. Edwin Slosson, professor of chemistry at the University of Wyoming, demonstrated this in 1899 when he stood before his students and poured a vial of distilled water over a ball of cotton wool. The water, he told them, was a highly aromatic chemical. He asked them to raise their hands when they could smell it. Within fifteen seconds most of the front row had their hands in the air. Forty-five seconds later, three-quarters of the audience were waving at him.

Slosson provided few details about his experiment, and so it barely rises above the level of anecdote. But on the evening of June 12, 1977, viewers of *Reports Extra*, a late-night British news show, became the unwitting subjects in a better-documented demonstration of the same phenomenon.

The show, which aired in Manchester, focused on the chemistry of sense. Toward the end of the program, viewers were shown a large cone with wires protruding from its point. The cone, they were told, represented a new form of technology—Raman spectroscopy. It would allow the station to transmit smells over the airwaves, from the studio straight into a viewer's living room.

The cone contained a "commonly known odorous sub-stance" that exuded a "pleasant country smell, not manure." The scent had been building up in the cone for the past twenty-three hours. Sensors were recording the vibrational frequencies of the odor molecules. These frequencies could then be broadcast over the air. Viewers' brains would interpret the frequencies as smells. *Voilà!* Smell-o-vision made real.

The station announced that an experimental smell trans-mission would occur in a few seconds. They asked viewers to report whatever they smelled, even if they smelled nothing at all. Then three, two, one . . . The screen changed to an oscil-loscope pattern and a tuning noise was heard. The smell had been transmitted.

Within the next twenty-four hours, the station received 179 responses. The highest number came from people who reported smelling hay or grass. Others reported their living rooms filling with the scent of flowers, lavender, apple blos-som, fruits, potatoes, and even homemade bread. A few smelled manure, despite this odor having been specifically excluded. Two people complained that the transmission brought on a severe bout of hay fever. Three others claimed the tone cleared their sinuses. Only sixteen people reported no smell sensation.

What is one to make of this? As far as the TV station knew, they had not actually beamed a smell over the airwaves, unless they accidentally did so by some unknown mechanism. The transmission was an experiment devised by Michael O'Mahony, a psychology lecturer at Bristol University (now at the University of California, Davis). He conceded that some respondents may have been lying, but assuming the majority told the truth, he offered what happened as a successful demonstration of the power of suggestion. He speculated that the suggestion worked either by causing people to imagine a nonexistent smell, or by prompting them to focus on a previously unnoticed odor in their environment.

Recent research indicates that suggestion not only influences *what* we smell, but also *how* we react to smells. In 2005 researchers at Oxford University asked subjects to sniff two odors, one labeled *cheddar cheese* and the other *body odor*. Predictably, the subjects rated the body odor as significantly more unpleasant. However, the two smells were identical. Only the labels differed. Consider that the next time you're enjoying some especially pungent cheese at a cocktail party.

O'Mahony, M. (1978). "Smell illusions and suggestions: Reports of smells contingent on tones played on television and radio." *Chemical Senses and Flavour* 3 (2): 183–89.

4. SIGHT

The Invisible Gorilla

You think it's going to be a test of your powers of concentration, and in a way it is. The researcher tells you he's going to show you a video of two teams, one dressed in black and the other in white, each throwing a basketball around. He asks you to count the number of times the white team passes the ball.

The video begins. The team members bob and weave. It's a little difficult to follow them. There are so many bodies moving around. But you think you're doing a pretty good job. You count the passes: one, two, three, four . . .

After about a minute the researcher stops the tape. You've got your number ready. "Fifteen passes," you tell him, but then he asks you something unexpected: "Did you notice the gorilla?"

Huh? You shake your head. What gorilla? "The gorilla that walked to the center of the screen, thumped its chest a few times, and then walked offscreen," the researcher replies. You look at him as though he's crazy. There was no gorilla in that video. But then he replays the tape and, sure enough, about halfway through it, a person in a gorilla suit does wander into the crowd of basketball players and thump her chest. How in the world did you miss that?

If you ever take the test described above, don't be surprised if you miss the gorilla. On average, 50 percent of first-time test takers fail to see it. Though, of course, now that you know the trick, you'll be looking for it. People who are simply asked to watch the video, with no further instructions, almost always see the creature.

Researchers Daniel Simons and Christopher Chabris conducted the gorilla experiment in 1998. They explain that so many people miss the hairy primate because of a visual phenomenon known as inattentional blindness. Our brains are only capable of focusing on a few details at any one time. We tune out everything else, literally becoming blind to it, even if we're staring right at it. So if something unexpected pops up—such as a woman in a gorilla suit—we fail to see it.

This explains why you might fail to notice friends waving at you in a crowded theater, because you're focusing on looking for an empty seat. Or why a pilot who's concentrating on an instrument display projected onto the windshield might not spot an unlooked-for plane crossing the runway up ahead. Perhaps it also explains why Bigfoot has eluded detection for so long. These unexpected "gorillas in our midst," as Simons and Chabris put it, are simply hard to see.

Incidentally, people who were asked to focus on the black-clothed team saw the gorilla far more often—83 percent of the time. Presumably this is because subjects who were visually tracking white-clothed players tuned out anything black, including black gorillas. If, instead, Ricardo Montalban had wandered by in a white suit, those who followed the black team probably would have missed him.

Perhaps the optical illusion Simons and Chabris revealed is merely an artifact of a lab setting. Surely in real life we would notice the gorilla. However, in the mid-1990s Simons and a different colleague, Daniel Levin, conducted an experiment that suggests not.

One of the researchers posed as a tourist seeking directions and approached random pedestrians on the campus of Cornell University. "Excuse me, do you know where Olin Library is?" he asked as he fumbled with a map. The researcher and the pedestrian conversed briefly, until workmen carrying a door suddenly barged between them. A moment later they resumed their conversation.

However, something had changed when the door passed between them. One of the workmen carrying the door was actually the second researcher, who surreptitiously switched places with his colleague and continued to converse with the pedestrian as though he had been there the entire time.

The two researchers were approximately the same age, but were dressed in different clothes. Amazingly, over half the subjects, eight out of fifteen, didn't realize they were talking to a new person until the researcher stopped them and asked, "Did you notice anything unusual at all when that door passed by a minute ago?" Many people replied, "Yes, those workmen were very rude." To which the researcher would reply, "Did you notice that I'm not the same person who approached you to ask for directions?" A bewildered look followed.

Like the invisible-gorilla test, the changing-tourist experiment reveals that we often fail to notice unexpected changes to an attended object. This phenomenon is known as "change blindness." Becoming aware of it can be rather unnerving. How much of the world around us might we be missing, one has to wonder. And can we ever again trust people who stop us and ask for directions on university campuses?

Of course, experiencing these effects for yourself is far more powerful than reading about them. If you go to Professor Simons's Web site, http://viscog.beckman.uiuc.edu/media/Boese.html, you can view videos of his research, including the experiments discussed above. However, you'll want to keep a close eye on the site. It has the potential to change at any time.

―――

Simons, D. J., & C. F. Chabris (1999). "Gorillas in our midst: Sustained inattentional blindness for dynamic events." *Perception* 28: 1059–74.

Through a Cat's Eyes

A young man stares at a movie screen. Restraints hold him in his chair. Small clamps keep his eyes pried open. He cannot blink. He must continue to watch as scene after scene of graphic violence plays on the screen.

Fans of Stanley Kubrick will recognize this scene from his movie *A Clockwork Orange*. The main character, Alex DeLarge, is subjected to a treatment called Ludovico aversion therapy, designed to transform him from an unruly thug into a non-violent, productive member of society. The treatment causes him to be crippled by nausea if he so much as thinks about violence, but it leads to tragic consequences. When released, Alex discovers he is powerless to defend himself against his numerous enemies, who duly take revenge on him.

Twenty-one years after the release of Kubrick's film, a strangely similar scene played out in a University of California laboratory—with one major difference. In Alex's place was an adult cat.

Researchers led by Dr. Yang Dan, an assistant professor of neurobiology, anesthetized a cat with Sodium Pentothal, chemically paralyzed it with Norcuron, and secured it tightly in a surgical frame. They then glued metal posts to the whites of its eyes, forcing it to look at a screen. Scene after scene played on the screen, but instead òf images of graphic violence, the cat had to watch something almost as terrifying—swaying trees and turtleneck-wearing men.

This was not a form of *Clockwork Orange*–style aversion therapy for cats. Instead, it was a remarkable attempt to tap into another creature's brain and see directly through its eyes. The researchers had inserted fiber electrodes into the vision-processing center of the cat's brain, a small group of cells called the lateral geniculate nucleus. The electrodes measured the electrical activity of the cells and transmitted this information to a nearby computer. Software then decoded the information and transformed it into a visual image.

The cat watched eight different short movies, and from the cat's brain the researchers extracted images that were very blurry, but were recognizably scenes from the movies. There were the trees, and there was that guy in the turtleneck. The researchers suggested that the picture quality could be improved in future experiments by measuring the activity of a larger number of brain cells.

The researchers had a purely scientific motive for the experiment. They hoped to gain insight into "the functions of neuronal circuits in sensory processing." But the commercial potential of the technology is mind-boggling. Imagine being able to see exactly what your cat is up to on its midnight prowls. Forget helmet cam at the Super Bowl; get ready for eye cam. Or how about this—never carry a camera again. Take pictures by blinking your eyes. It would work great unless you had a few too many drinks on vacation!

Dan, Y., et al. (1999). "Reconstruction of Natural Scenes from Ensemble Responses in the Lateral Geniculate Nucleus." *Journal of Neuroscience* 19 (18): 8036–42.

5. SOUND

The Mozart Effect

Mozart has a new hangout. No longer relegated to the dusty stereos of classical-music buffs, he can now be heard drowning out the sounds of crying babies and squealing Teletubbies at the local nursery, or blasting from high-end toys and crib mobiles.

Why has Mozart become so popular with the under-five set? The reason traces back to the startling results of a 1993 experiment performed by Frances Rauscher, Gordon Shaw, and Katherine Ky at the University of California, Irvine.

In the experiment, thirty-six college students each tried to solve three sets of spatial-reasoning tasks. A typical task consisted of imagining a piece of paper folded and cut in various ways, and then figuring out what the paper would look like unfolded. Before starting each new task, the subjects sat through a different ten-minute "listening condition." Before the first task they heard Mozart's Sonata for Two Pianos in D Major, K. 448; before the second one, a blood-pressure-lowering relaxation tape; and prior to the third, silence.

The final results showed a clear trend. The students scored highest on the task after listening to Mozart. In fact, the improvement was quite dramatic: "The IQs of subjects participating in the music condition were 8–9 points above their IQ scores in the other two conditions."

Consider the implications of this. An almost ten-point leap in IQ—albeit a temporary one, as the effect seemed to fade away after fifteen minutes—just by listening to Mozart. Getting smart fast had never been so easy.

These results got people's attention. Soon the "Mozart Effect," as it came to be known, was being tested in all kinds of situations.

High school students started playing Mozart while studying for exams. University of Texas scientists combined Mozart's music with whole-body vibrotactile stimulation to see if this would enhance the effect—it didn't. A Texas prison made inmates listen to the composer during classes. Rauscher, in a follow-up experiment, claimed Mozart's music improved maze learning in rats. Korean gardeners declared they had a season of spectacular blossoms after playing Mozart in a field of roses. Finnish researchers looked into whether the effect improved the memory skills of monkeys. To their surprise, Mozart actually had a negative effect. Our primate cousins evidently aren't classical-music lovers.

Scientists also considered the work of other composers. The original researchers theorized that the complexity of Mozart's music somehow stimulated neurons in the cortex of the brain. They explained that "we chose Mozart since he was composing at the age of four. Thus we expect that Mozart was exploiting the inherent repertoire of spatial-temporal firing patterns in the cortex." The work of other "complex" musicians—such as Schubert, Mendelssohn, and, of all people, Yanni—was found to share Mozart's enhancing properties. Noncomplex, unenhancing musicians included Philip Glass, Pearl Jam, and Alice in Chains.

But popular interest in the phenomenon didn't really explode until word got out that a little bit of Mozart could increase infant intelligence. Ambitious parents, eager to have a junior genius, promptly went Mozart mad. Mozart-for-baby

CDs rocketed to the top of the charts. The sounds of the composer began to be piped into nurseries. Zell Miller, governor of Georgia, ordered that Mozart CDs be distributed to all infants born in the state, and the state of Florida passed a law requiring classical music be played in state-funded day-care centers.

The curious thing was that not a single experiment had ever suggested a link between listening to Mozart's music and increased infant intelligence. The closest an experiment had come to making this connection was a 1997 study, again by Rauscher, that demonstrated a relationship between piano lessons and improved spatial-reasoning skills among preschoolers. Learning to play the piano, of course, is not the same as listening to a CD.

The massive amount of popular interest in the Mozart Effect prompted the scientific community to take a closer look at the phenomenon. That's when the theory began to hit rocky ground. Many researchers reported a failure to replicate the results of the 1993 study. In response, the UC Irvine team clarified that Mozart's music did not appear to have an effect on all forms of IQ, but rather on spatial-temporal IQ, the kind that applied to paper folding and cutting tasks. In other words, millions of parents were unwittingly priming their children to become master scrapbookers. But even with this narrower focus, other labs continued to report a failure to replicate the results.

In 1999 and 2000 two researchers, Christopher Chabris (whom we just met in the invisible-gorilla experiment) and Lois Hetland, separately analyzed all the experimental data on the Mozart Effect. Both concluded that while a temporary effect did appear to exist, it was negligible. As for its application to children, Hetland bluntly dismissed this:

The existence of a short-lived effect by which music enhances spatial-temporal performance in adults does not lead to the conclusion that exposing children to classical music will raise their intelligence or their academic achievement or even their long-term spatial skills.

These negative results threw some cold water on the Mozart Effect's scientific credibility, but they hardly dimmed its mass-market popularity. A sprawling self-help industry continues to promote the benefits of Mozart via books, CDs, Web sites, and countless baby toys. One entrepreneur, who has trademarked the term *Mozart Effect*, even claims the composer's music has medical benefits. He tells how he dissolved a large blood clot behind his eye by humming it away. That sound you hear now is Mozart turning over in his grave.

Rauscher, F. H., G. L. Shaw, & K. N. Ky (1993). "Music and Spatial Task Performance." *Nature* 365: 611.

The Acoustics of Cocktail Parties

With a drink gripped precariously in one hand, you lean closer toward your fellow guest. "What did you say?"—"I really don't know what she sees in him."—"Beg your pardon?"—You lean closer still. More people are arriving fast. The background murmur of voices is rising to a din. It's growing harder by the minute to hear anyone at this cocktail party. "I said, I REALLY DON'T KNOW WHY SHE GOES OUT WITH HIM."

In January 1959 William MacLean theorized in the *Journal of the Acoustical Society of America* that as any cocktail party grows in size there will arrive a moment when the gathering abruptly ceases to be quiet and becomes loud. This is the moment when guests start to crowd together and raise their voices to be heard above the background noise. He produced a mathematical formula to predict exactly when this would occur:

$$N < N_0 = K \left[1 + \frac{(aV/4\pi h) + d_0^2}{d_0^2 S_m^2} \right]$$

In this formula N is the number of guests at the party, K the number of guests per conversational group, a the sound absorption coefficient of the room, V the room volume, h the "properly weighted mean free path . . . of a ray of sound through the room," d_0 the minimum conventional distance between talkers, and S_m the minimum signal-to-noise ratio required for intelligible conversation. To calculate the maximum number of guests compatible with a quiet party, solve for N_0. Quite simple, really.

Acoustical researchers were not about to let MacLean's hypothesis go untested. Throughout the remainder of 1959 the staff of the Division of Building Research of the National Research Council of Canada fanned out at cocktail parties, acoustical equipment in hand, to gather experimental evidence that would either confirm or refute MacLean's theory.

The investigators admitted a few of the gatherings they monitored may have seemed like regrettable choices, in light of the nature of the research. For instance, collecting data at a cocktail party attended entirely by librarians, "a group dedicated professionally to maintaining quiet," would seem to defeat the purpose. However, they insisted the librarians were actually quite raucous.

MacLean's theory predicted "an abrupt 15-dB transition at the critical point." This was not experimentally confirmed.

Instead, noise levels rose steadily. There was no evidence of an abrupt transition point. The NRC scientists suggested MacLean's formula failed to factor in phenomena such as appreciative laughter and wide variability in the speech power of talkers.

Of course, this entire line of research assumed cocktail parties populated by well-mannered guests who did not have to compete with blaring music. At parties where music is blasting, such as the typical college party, the research of Charles Lebo, Kenward Oliphant, and John Garrett would be of more use. During the 1960s these three doctors investigated acoustic trauma from rock-and-roll music by measuring sound levels at "typical San Francisco Bay Area rock-and-roll establishments frequented almost exclusively by teenagers and young adults, of whom many fall into a group popularly designated as 'hippies.'" They discovered sound levels far in excess of those considered safe. They made the following suggestion to the hippies:

> Attenuation of the amplification to safe levels would substantially reduce the risk of ear injury in the audience and performers and, in the opinion of the authors, would still permit enjoyment of the musical material.

The hippies would have heeded the warning—really, they would have—but the music was too loud to hear what the doctors were saying.

Legget, R. F., & T. D. Northwood (1960). "Noise Surveys of Cocktail Parties." *Journal of the Acoustical Society of America* 32 (1): 16–18.

CHAPTER THREE

Total Recall

Memory, the theme of this chapter, is an ancient obsession. For millennia people have tried to find ways to increase it, delete it, or hold on to what they have. In the sixteenth century, an Italian inventor named Giulio Camillo claimed to have designed a Theater of Memory that enabled scholars to memorize all forms of knowledge, in their entirety. The Memory Theater was supposed to be a physical structure, although whether it was ever built or merely existed as plans on paper is not known. A scholar would stand inside the theater and see arrayed before him tiers of wooden shelves bearing cryptic images, each of which represented a form of knowledge. Studying the location and meaning of these images, Camillo claimed, would allow vast amounts of information to somehow, magically, flow into the savant's brain. Needless to say, there is no evidence this worked. In the modern world, Hollywood has envisioned equally fantastic memory-altering technologies. For instance, there was the Neuralizer, carried by the government agents in the *Men in Black* films, that erased the memory of anyone who stared into its flash; or the Rekall mind-device machine, from the

Arnold Schwarzenegger movie *Total Recall*, that allowed people to take imaginary adventures by implanting false memories of what they had done. In real life, scientific researchers have not yet achieved the kind of total memory control artists have dreamed of, but not for lack of trying.

Electric Recall

Wilder Penfield is poking around in your head—literally. You lie in an operating room. The top of your skull has been cut away, revealing your brain. But you are still awake. If there was a mirror on the ceiling, you could see the Canadian neurosurgeon moving behind you. Penfield lifts up an instrument, a monopolar silver-ball electrode, and touches it to your brain. You cannot feel this, because there are no nerve endings in the brain. But suddenly a memory flashes before your eyes, something you hadn't thought of in years. You see your mother and father standing in the living room of the house you grew up in, and they're singing. You listen closely. It's a Christmas carol. "Deck the halls with boughs of holly, fa-la-la-la-la, la-la-la-la." The tune is so clear you can hum along with it. You start to do this, but just then Dr. Penfield removes the electrode, and the memory vanishes as quickly as it appeared.

The phenomenon you have just experienced is electric recall. While performing brain surgery on epileptic patients during the 1930s and 1940s at the Montreal Neurological Institute, Penfield discovered that sometimes, when he touched an electrode to their brains, random memories would intrude into their conscious thoughts. It was as though he had found the mind's videotape archive. When he pushed the magic button, *zap*, scenes from the patients' pasts would start playing. Penfield himself used this analogy: "Applying the stimulus was like pressing the start button on a tape recorder. Memories would start playing before the patient's eyes, in real time."

Penfield was poking around in these brains to orient himself during the surgical procedure—because everyone's neurons are wired a little differently—as well as to locate damaged regions. He would touch his electrode directly to a region, such as a wrinkle on the temporal lobe, and ask the patient what sensation, if any, she felt. Then he would stick a numbered piece of paper on that spot. When he was finished he took a picture of all the little pieces of paper. The resulting photo served as a convenient map of the patient's brain he could then refer to as he worked. Kind of like surgery by numbers.

The first time one of his patients reported spontaneous memory recall was in 1931. He was operating on a thirty-seven-year-old housewife. When he stimulated her temporal lobe with an electrode, she suddenly said that she "seemed to see herself giving birth to her baby girl."

Penfield was sure he had stumbled upon evidence of a memory library within the brain. He imagined it as "a permanent record of the stream of consciousness; a record that is much more complete and detailed than the memories that any man can recall by voluntary effort." He began a systematic search for this memory library in other patients. Over a period of more than twenty years, he touched his electrode to hundreds of exposed brains, prompting subjects to report a variety of memories. These memories included "watching a guy crawl through a hole in the fence at a base-ball game"; "standing on the corner of Jacob and Washington, South Bend, Indiana"; "grabbing a stick out of a dog's mouth";

"watching a man fighting"; "standing in the bathroom at school"; and "watching circus wagons one night years ago in childhood."

It was as though Penfield were Albus Dumbledore of *Harry Potter* fame, dipping his magic wand into a Pensieve and pulling out stray, glittering thoughts. The science fiction quality of all this was not lost on author Philip K. Dick. In his novel *Do Androids Dream of Electric Sheep?*, later adapted for the screen as *Blade Runner*, characters use a device called a Penfield Mood Organ to dial up emotions on command. (Dick also wrote the novel on which the movie *Total Recall* was based.)

Penfield's discovery generated excitement during the 1950s, when he first publicly revealed his findings. Some hailed it as clinical confirmation of the psychoanalytic concept of repressed memories. But as time wore on, the scientific community grew more skeptical. Other neurosurgeons failed to replicate Penfield's results. In 1971 doctors Paul Fedio and John Van Buren of the National Institute of Neurological Diseases and Stroke in Maryland stated bluntly that, in their extensive work with epileptic patients, they had never witnessed the phenomenon Penfield had reported. Brain researchers noted that it was definitely possible to provoke brief hallucinations by means of electrical stimulation of the brain, and experimenters such as Elizabeth Loftus of UC Irvine, whom we shall meet again in a few pages, built on this observation to argue that Penfield must have mistaken such hallucinations for memories. Basically, not many brain scientists today take seriously Penfield's idea of a complete memory library hidden in our brain.

Still, it would be cool if Penfield were right and we *could* access everything we had ever seen or heard. We could press a button on a remote control and remember where we parked our car, or what we were supposed to buy at the supermarket. The only problem is that we'd still end up forgetting where we put the remote.

Penfield, W., & P. Perot (1963). "The brain's record of auditory and visual experience." *Brain* 86: 595–696.

Elephants Never Forget

An elephant walks into a bar and challenges the bartender to a memory contest. "Loser pays for the drinks," says the elephant. What should the bartender do?

Before answering, the bartender might want to consider the elephant-memory experiments of Bernhard Rensch. During the 1950s Rensch explored the relationship between brain size and intelligence in the animal kingdom. This led him to conduct a series of tests on a five-year-old Indian elephant at the Münster Zoological Institute, of which he was the director. His results suggested that while it's not literally true that elephants never forget, they do have excellent memories.

Rensch started with a simple test. He presented the elephant (never identified by name) with two boxes, each marked by a different pattern, a cross or a circle. Would she remember that the box with the cross on its lid always contained food? It took her a while, over 330 tries, but eventually she figured it out. Rensch helped her by screaming "nein!" every time she chose the wrong box. Once she got the idea, she really got the idea. From then on she consistently chose the cross over the circle.

Rensch next introduced her to new pairs of positive (food) and negative (no food) patterns: stripes, curvy lines, dots, etc. Now that she understood the game, she was on a roll. She quickly mastered twenty pairs, a total of forty symbols. In a test using all the symbols, given in random order, she chose the correct pattern almost six hundred times in a row. Many humans would be hard-pressed to do as well.

The elephant could also pick out the correct box from a choice of three negative patterns and a positive one. However, when Rensch presented her with a negative-patterned box and a box with a blank lid (neither of which contained food), she got mad, tore the lids off the boxes, and trampled them. Apparently, elephants don't like trick questions.

The hardest test was yet to come. Rensch waited a full year and showed the elephant thirteen of the symbol pairs she had previously learned. She immediately recognized them. In 520 successive trials, she scored between 73 and 100 percent on all the pairs except one, a double circle versus a double half circle. And even on that pair she scored 67 percent. Rensch declared it to be "a truly impressive scientific demonstration of the adage that 'elephants never forget.'"

Science can't generalize about an entire species based on a sample of one. Perhaps Rensch's subject happened to be a genius. However, similar tests have confirmed the remarkable recall of elephants.

In 1964 Leslie Squier trained three elephants at the Portland Zoo to distinguish between lights of different color. They received a sugar cube as a reward for a correct response. Eight years later Hal Markowitz salvaged Squier's equipment from a scrap heap and retested the elephants. One of them, Tuy Hoa, walked right up and gave the correct answers. She clearly remembered the test. The other two elephants didn't perform as well. But when Markowitz examined them he realized there was a reason for this. They were almost blind and couldn't see the lights.

Given all this, how does our bartender respond to the challenge? Simple. He throws the elephant out on its trunk.

> The elephant: "Why did you do that?"
> The bartender: "Because you never paid your bill last time you were in here."
> The elephant: "That was three years ago. I didn't think you would remember."
> The moral: Bartenders never forget, either.

Rensch, B. (1957). "The Intelligence of Elephants." *Scientific American* 196 (2): 44–49.

The Memory Skills of Cocktail Waitresses

Barmaids, it turns out, have pretty good memories, too. Anecdotal accounts have them remembering up to fifty drink orders at once on busy nights. Suspecting cocktail waitresses might be a previously unrecognized population of master mnemonists, Professor Henry Bennett of the University of California, Davis, set out to test just how good their memory skills were.

During the early 1980s he and a coinvestigator canvassed bars in San Francisco and Sacramento searching for waitresses willing to participate in their experiment. Whenever they found one who was agreeable, they whipped out a portable testing kit—a Ken-and-Barbie-style cocktail lounge housed in a suitcase. It had two miniature tables (covered with green felt), chairs, and male and female dolls, aka customers. The dolls were decorated "with different fabrics and jewellery appropriate to the doll's gender. Hair color was painted on and some males received beards and/or moustaches so that, as in real life, each doll customer was a unique individual." Bennett did not record whether the dolls wore '80s-appropriate fashions such as Jordache jeans or *Flashdance*-style leggings.

As the life-size waitress looked on, the miniature customers placed their orders. They proceeded either sequentially around the table or in random order. A tape recording supplied their voices: *Bring me a margarita . . . I'll have a Budweiser.* The experimenter wiggled each doll in turn to indicate who was speaking. Following a two-minute waiting period, the waitress had to deliver the drinks, which were "small laboratory rubber stoppers" with drink name flags poked into them. No real alcohol for Ken and Barbie.

After forty waitresses had been tested, Bennett repeated the experiment on forty UC Davis undergraduates. The result: The waitresses wiped the floor with the students. They averaged 90 percent correct, versus the students' 77 percent. Furthermore, "waitresses were nearly twice as efficient in time to place each drink as were students."

When interviewed, the waitresses had no idea why they were so good at remembering drink orders. They had never received any special training. It was just a skill picked up on the job. Intriguingly, most of them commented that their memory skills improved on busy nights when they got "in the flow."

This stands in stark contrast to the memory skills of bar patrons. Their powers of recall decline on busy nights in an almost perfectly inverse relationship to the improvement seen in barmaids.

———

Bennett, H. L. (1983). "Remembering Drink Orders: The Memory Skills of Cocktail Waitresses." *Human Learning* 2: 157–69.

Underwater Memory

Researchers often choose unusual locations to stage their experiments. We have already encountered memory experiments conducted in an operating room, a zoo, and a bar. But the prize for the most unusual setting for research in this field goes to Duncan Godden and Alan Baddeley of the University of Stirling. They tested the memories of subjects who were submerged twenty-feet deep in the chill Atlantic waters off the coast of Scotland.

The year was 1975. Eight divers from the Scottish Sub-Aqua Club descended to the bottom of Oban Bay, where they sat anchored down by weights, holding in their hands a Formica board and pencil. Eight other divers, also dressed in scuba gear, remained on the beach. All participants then heard through their headsets a list of words. The list was read twice. The question Godden and Baddeley sought to answer was this: Would the words learned underwater be best recalled underwater?

This may not seem like a question of great practical value, except for those planning to take an exam in a swimming pool. But as often is the case with such things, there was more to it than met the eye. What Godden and Baddeley were really interested in was context dependency in memory. Is there a link between memory and place? For instance, if you learn a subject in a particular classroom, is it easier to recall that subject in the same classroom?

After hearing the list of words, four of the submerged divers ascended, and four of the divers on the beach dived into the water. All sixteen participants then wrote down as many of the words as they could remember. The results were clear. Those who remained in the same place, whether underwater or on land, scored higher than those who moved. Environment appeared to have a large impact on recall.

Could the effort of moving have disrupted attempts at memorization? To address this possibility, the experimenters ran a second test in which, in between hearing the word list and taking the test, the subjects on the beach briefly dived into the water and returned to the beach. The disruption had no significant effect on memorization.

Due to the Scottish climate, this research involved some hardship. Godden and Baddeley noted that "subjects began each session in roughly the same state, that is, wet and cold." There was also risk. While underwater one diver almost got run over by an amphibious army truck that happened to be passing by. But it was worth it. The experiment was well received and is frequently cited as evidence of a same-context advantage in learning.

Nevertheless, recent studies have cast doubt on whether Godden and Baddeley's results can be generalized to other contexts. A 2003 Utrecht University study repeated their experiment in a more down-to-earth setting—the university's medical center. Sixty-three medical students were given lists of words and patient case descriptions to learn in both a clinical bedside setting and a classroom. No same-context advantage was observed. Of course, the Utrecht study did not require the medical students to wear full scuba gear. This may have played a role in the differing results.

———

Godden, D. R., & A. D. Baddeley (1975). "Context-dependent memory in two natural environments: On land and underwater." *British Journal of Psychology* 66 (3): 325–31.

Edible Memory

A mother takes her child to the doctor to get his shots. As the doctor prepares the usual measles and tetanus injections, he inquires casually, "Would you be interested in any French or Spanish for Johnny?" The mother considers a second. "Yes, I think he should know French. Why don't you give him a shot of that. Oh, and I noticed he was making some mistakes in his piano practice last week. Could you also give him a booster of piano?"

Might acquiring a new skill someday be as easy as going to a doctor's office and getting an injection? It's not likely, but between the 1950s and 1970s, there was a brief, heady period when many thought it was a real possibility. In 1959 *Newsweek* declared, "It may be that in the schools of the future students will facilitate the ability to retain information with chemical injections." In 1964 the *Saturday Evening Post* looked forward to a day when people would be able to "learn the piano by taking a pill, or to take calculus by injection." The cause of this excitement was a series of unusual experiments involving, of all things, cannibalistic flatworms.

The experiments began in 1953 at the University of Texas, where James McConnell and Robert Thompson were students.

They were interested in what was, at the time, a relatively new theory of memory—the synaptic theory, which posited that memories form when new synaptic connections are made between neurons in the brain. Fate was set in motion when the two decided to use flatworms to study this theory.

Flatworms, also known as planarians, are humble little creatures. They measure only half an inch to two inches in size. They have pointed heads, tiny eye spots, and slimy, tubelike bodies. There are millions of them around, lurking beneath rocks near ponds and rivers, but most humans never notice them. McConnell and Thompson singled them out as research subjects because flatworms are just about the simplest forms of life to have neurons and synapses. But flatworms don't have many neurons, which, it was hoped, would simplify matters from a research perspective.

The big question was, could flatworms learn? If not, they would be useless as subjects of memory experiments. Most scientists assumed they couldn't learn, but McConnell and Thompson set out to prove the prevailing wisdom about planarians wrong.

Literally working out of their kitchen sink at first, the two students designed a flatworm training device. It consisted of a shallow, water-filled trough into which they placed a single worm. Over the trough they positioned a pair of bright lights. They would turn on the lights for two or three seconds, and then give the worm a jolt of electricity, causing its body to scrunch up. They repeated this procedure over a hundred times until eventually the tiny creature scrunched up as soon as the lights went on, in anticipation of the shock. This hardly made the worm the intellectual equivalent of Einstein, but it did demonstrate the worm had learned something—to associate the light with the imminent arrival of a shock.

Rival researchers found the idea of trained flatworms odd, but believable. If the flatworm research had ended there it would be remembered today as a solid piece of scientific work. However, it didn't end there. In 1956 McConnell moved to the University of Michigan, where he continued to study flatworms. His experiments there would briefly make the worms international celebrities and start magazines speculating about memory pills and injectable learning.

One of the interesting things about flatworms is that if you cut them in half, each half—the head and the tail—will regenerate a complete body in about two weeks. It's an enviable skill. McConnell decided to see what would happen if he trained some worms and then cut them in half. Would each half, once regenerated, retain the knowledge of the original worm? He anticipated the head would, but doubted the tail's abilities. After all, there's no brain in the tail. But to his surprise, the tails did appear to retain learning. Or to be more specific, the tails relearned the scrunching trick far more quickly than naive planarians (i.e., planarians that had never been exposed to training), which implied the tails had some memory of what they had learned. In fact, the tails performed better than the heads.

The scientific community greeted these results with deep skepticism. What McConnell was claiming was somewhat like saying memories could be transferred from one person to another via a leg transplant. It just didn't seem possible, and it completely contradicted the synaptic theory of learning.

However, McConnell felt sure his results were correct. He decided it was the synaptic theory that was wrong, and he developed a rival theory of memory based on his findings. According to his theory, memories were not formed by neurons making new connections between one another, but rather were encoded onto molecules inside cells. He speculated that this "memory molecule" might be ribonucleic acid (RNA), the biochemical cousin of DNA.

Again, the mainstream scientific community had a tough time swallowing McConnell's theory, and not just because his claims were so extraordinary. Many felt he was too much of a clown to be taken seriously.

McConnell seemed to go out of his way to tweak the sensibilities of the scientific establishment. He loved making grand, far-reaching claims about his research to the media —claims that annoyed his rivals. As interest in flatworm studies spread and a small community of memory-transfer investigators emerged, he started publishing a journal called the *Worm-Runner's Digest*, to serve as a clearinghouse for the latest research, but it was hardly a typical journal. Instead, imagine a combination of *Mad Magazine* and the *Journal of Neuroscience*. The cover sported a two-headed flatworm coat of arms. Inside, serious articles rubbed shoulders with satire, comic verse, and cartoons. Under pressure from his contributors, McConnell eventually changed the format so that the serious content occupied the first half (which he renamed the *Journal of Biological Psychology*), and the humorous stuff —printed upside down—the second half. Which, in a way, made the journal even weirder. One had to flip it around to read the entire thing.

And yet, once again, had McConnell stopped there, his worm studies would still be remembered as intriguing (though unorthodox) work. But he didn't stop, and it was his next experiment that not only made him famous, but also convinced his critics that he had traveled irrevocably into the land of woo-woo.

McConnell wanted to test his RNA hypothesis. He imagined the best way to do so would be to extract RNA from a trained worm, inject it into a naive worm, and then observe whether memory had been transferred. But he had no good way of extracting pure RNA from a worm. So, taking advantage of a second interesting thing about flatworms—that they happen to be cannibalistic—he settled on a far cruder method. McConnell simply chopped up a trained worm and fed it to a naive worm. He wrote:

> Once the cannibals had had several meals of "educated" tissue, they were given their first conditioning sessions. To our great surprise (and pleasure), from their very first trials onward the cannibals showed significant evidence that they had somehow "ingested" part of the training along with the trained tissue . . . Somehow, in a fashion still not clear to us, some part of the learning process seems to be transferable from one flatworm to another via ingestion.

Whereas his previous work had been met with polite skepticism, these new claims evoked howls of disbelief. The general reaction of the research community—apart from McConnell's small band of fellow memory-transfer enthusiasts—was something along the lines of, "You've got to be kidding!" It was like suggesting someone could acquire all of Einstein's knowledge by eating his brain. Of course, McConnell was quick to point out that though RNA could survive a journey through a flatworm's simple digestive system, it wouldn't last long in the harsh, acidic environment of a human stomach—meaning that, even if his hypothesis was correct, RNA transfer wouldn't work in humans via ingestion. However, the media overlooked this point and happily fantasized about a coming age of edible memory.

The debate over the biochemical transfer of memory dragged on for years. Many labs investigated the phenomenon. Some got positive results and supported McConnell's hypothesis, but far more got negative results. Critics suggested the positive results were caused by experimental bias—the researchers were seeing what they wanted to see by misinterpreting the behavior of the flatworms. Supporters upped the ante by claiming to have found evidence of memory transfer in mammals such as rats. Critics fired back with a letter to the journal *Science*, in which twenty-three different researchers stated they had found no evidence that biochemical memory transfer worked in rats. Defenders of the theory insisted the critics had conducted poorly designed experiments.

However, with each round of the battle the supporters of memory transfer grew weaker. Funding began to dry up. The broader scientific community lost interest and turned its attention elsewhere. The memory-transfer theory was dying by attrition. At last even McConnell moved on to other things, leaving one lone enthusiast, a Baylor College of Medicine researcher named Georges Ungar, to fight the battle.

Ungar was convinced he had found evidence of memory transfer in rats. He had trained rats to fear the dark, then ground up their brains and injected extracts into the brains of untrained rats, who subsequently seemed to show a similar fear of the dark. To satisfy his critics, he decided to isolate what he believed to be the memory molecule from the brains of these rats in quantities suitable for widespread analysis. But training thousands of rats and grinding up their brains was a costly, labor-intensive procedure. He soon realized this was impractical. He needed an animal that could be more cheaply trained and dissected in mass quantities, and so he hit upon the idea of using the goldfish. Ungar announced a plan to train thirty thousand goldfish to fear colored lights. He would then decapitate the fish and remove their brains, creating a two-pound stockpile of the memory substance. Unfortunately, Ungar's old age and death prevented the fulfillment of his scheme. For this, the goldfish were thankful.

With the death of Ungar, memory transfer lost its last great champion. And though the debate over biochemical transfer of memory went on for almost two decades, today you're unlikely to find it mentioned in many textbooks. It's as though it never happened, gone from the collective memory of science like a case of mild indigestion that caused brief discomfort, and was soon forgotten.

McConnell, J. V. (1964). "Cannibalism and memory in flatworms." *New Scientist* 21 (379): 465–68.

Beneficial Brainwashing

Mary C. checked into a clinic complaining of menopause-related anxiety. She probably imagined a few weeks of rest and relaxation awaited her, perhaps some psychological counseling. She couldn't have anticipated what actually lay in store. First came the massive doses of LSD. Next was the intensive electroshock therapy. Soon she had no memory of her past. She didn't even know her own name. She stumbled blindly through the hallways of the clinic, drooling and incontinent. But there was more to come—thirty-five days locked inside a sensory-deprivation chamber, topped off by three months of drugged sleep as a tape-recorded voice spoke the same phrases over and over from speakers inside her pillow: *People like you and need you. You have confidence in yourself.*

Mary C. had the misfortune of coming under the care of Dr. Ewen Cameron, director of the Allan Memorial Clinic in Montreal, Canada. She became one of the hundreds of unwitting subjects in his CIA-funded "beneficial brainwashing" experiments.

Born in Scotland, the son of a Presbyterian minister, Cameron had clawed his way to the top, fueled by fierce ambition. By the late 1950s he was one of the most respected psychiatrists in the world. He had served as president of the Quebec, Canadian, and American Psychiatric Associations and would go on to cofound the World Psychiatric Association. But it vexed him that one treasure eluded his grasp—the Nobel Prize. He embarked on a program of experimentation to discover a cure for schizophrenia, sure this would net him the trophy. His patients served as his subjects, whether they wished to or not, and often whether they had schizophrenia or not.

The cure that Cameron dreamed up was a testament to his hubris. He haphazardly cobbled it together from bits and pieces of other experimental therapies. Whatever caught his eye, he tried, constructing a true Frankenstein's creation.

The premise of his cure was to wipe schizophrenic patients' minds clean, erasing all their memories, and then to insert new nonschizophrenic personalities into their empty brains. Drawing on Cold War imagery, he described this as beneficial brainwashing, capable of transforming the mentally ill into healthy *new* people. A modern analogy would be fixing a software error in a computer by erasing everything on the hard drive and installing an entirely new version of the operating system. Except, of course, the human brain is not like a computer. It cannot simply be erased and reformatted.

Step one was the Mind Wipe. Cameron euphemistically

referred to this as "depatterning." He stripped away the mind's defenses and memories using electroshock therapy. Electroshock was widely used during the 1950s, but Cameron applied it far more aggressively than most doctors dared, administering it multiple times a day, at levels six times the normal charge. In essence, he fried his patients' brains. And for good measure he topped this off with massive doses of mind-altering drugs such as LSD.

Reducing patients to walking zombies may seem like the kind of thing most physicians would not own up to in public, but remember that Cameron fully believed he was in the running for a Nobel Prize. So he voluntarily published the details. You can go to your local library and read about it in his own words. In the following description, taken from a 1960 article in the journal *Comprehensive Psychiatry*, Cameron itemized the effects of depatterning on one unlucky patient:

> The patient loses all recollection of the fact that he formerly possessed a space-time image which served to explain the events of the day to him. With this loss, all anxiety also disappears. In the third stage, his conceptual span is limited to a few minutes and to entirely concrete events. He volunteers a few statements on questioning: He says he is sleepy or that he feels fine. He cannot conceptualize where he is, nor does he recognize those who treat him . . . What the patient talks about are only his sensations of the moment, and he talks about them almost exclusively in highly concrete terms. His remarks are entirely uninfluenced by previous recollections—nor are they governed in any way by his forward anticipations. He lives in the immediate present. All schizophrenic symptoms have disappeared. There is complete amnesia for all events of his life.

Having destroyed the mind, the next stage was to rebuild it. The hope was that some of the patient's memories would spontaneously return as he recovered from the electroshock. Sometimes they did. Sometimes they didn't. Meanwhile, Cameron labored to impose healthy thought patterns in place of the unhealthy, schizophrenic ones. To achieve this he used a process he called "psychic driving."

The concept of psychic driving occurred to Cameron after he read about an American inventor, Max Sherover, who had created a sleep-learning machine—a modified record player that repeated messages as a person dreamed. The sleeping mind supposedly soaked up these messages, and the person woke knowing a foreign language or how to send a signal in Morse code. (We'll encounter sleep learning again in chapter four.)

Cameron thought a modified version of this might be used to reprogram a patient's mind. He experimented with the technique on its own, before pairing it with depatterning. He made neurotic patients wear headphones and listen to anxiety-provoking statements, in an attempt to force them to confront their fears. The statements repeated over and over until the patients couldn't stand it any longer. Often they grew so enraged they flung the headphones across the room and stormed out the door.

Again, Cameron touted this in public as a cutting-edge therapy. He even invited the press to the clinic to witness a patient undergoing psychic driving. In the resulting article in Montreal's *Weekend Magazine* in 1955, we find a picture of a woman on a bed squirming in discomfort while listening to a looping message. One of Cameron's assistants stares at her impassively as he fiddles with a knob on the tape player.

Cameron's ambitions for the uses of psychic driving quickly grew by leaps and bounds. He didn't want patients merely to confront their anxieties—he wanted to violently drive messages into their psyche to reshape their very identities. He imagined exposing patients to a message all day for months on end. Of course, patients resisted listening for even a few minutes, so they weren't about to lie still voluntarily as a message repeated hundreds of thousands of times. Therefore, Cameron devised ways to force them to listen. He injected patients with curare, a paralyzing South American plant toxin, compelling them to lie motionless in bed as a voice droned from speakers in their pillows. Or he drugged them with barbiturates, sending them into a deep sleep for weeks at a time as the tapes whispered in their ears. Or he confined them in a sensory-deprivation chamber—a soundproof cubicle—with goggles over their eyes, and restraints holding their arms in place.

Cameron encountered some unexpected problems. The engineer he hired to record some of the messages spoke with a strong Polish accent, and patients subsequently reported their thoughts were taking on a Polish cadence. Cameron corrected this by rerecording the messages in his own voice. Since he had a thick Scottish burr, this was probably not an improvement.

Patients were also prone to misunderstanding the message. According to one oft-repeated rumor, a woman came to Cameron complaining of an inability to feel sexually comfortable with her husband. On Cameron's orders, she lay in bed for weeks listening to the phrase "Jane, you are at ease with your husband." Only later did Cameron discover she thought the message was saying "Jane, you are a tease with your husband."

Overall Cameron thought psychic driving showed great promise. In one test of the technique, he placed patients into a drugged sleep and made them listen to the message, "When you see a piece of paper, you want to pick it up." Later he drove them to a local gymnasium. There, lying in the middle of the gym floor, was a single piece of paper. Reportedly many of them walked over to pick it up.

Thanks to his public boasts about the promise of beneficial brainwashing, Cameron came to the attention of the Central Intelligence Agency. If he was going to develop a workable brainwashing technique, they wanted to know about it. In 1957 the agency began surreptitiously funneling money to him via one of its front organizations, the Society for the Investigation of Human Ecology. But after a few years, despite Cameron's enthusiastic progress reports, the agency realized his process simply didn't work. They summarily ended his funding in 1961. Soon after, Cameron grudgingly conceded his experiments had been "a ten year trip down the wrong road." Of course, he never bothered to apologize to the patients whose lives he had ruined. He did, however, shred their hospital records.

The CIA later came to regret getting involved with Cameron. Details of the funding arrangement leaked out in the late 1970s, leading to a lawsuit filed against the agency by a group of Cameron's former patients. The suit was settled out of court for an undisclosed amount.

Cameron spent the remainder of his career working on a project that was, in many ways, the polar opposite of his brainwashing experiments. Instead of destroying memory, he attempted to restore it. He had read about James McConnell's memory-transfer experiments, and with his keen talent for attaching himself to questionable research, he eagerly jumped on this bandwagon. Once again, visions of a Nobel Prize danced before his eyes.

McConnell had theorized that RNA played a key role in the process of memory formation, so Cameron gave elderly patients RNA pills, hopeful this would improve their recall. He periodically administered a memory test to check their progress. Soon he was reporting positive results. The only problem was that he was giving the patients the exact same test, again and again, which made their improvement predictable and rather meaningless. Somehow he had overlooked this obvious flaw in his experiment's design.

In September 1967 Cameron climbed to the summit of a mountain near Lake Placid and dropped dead of a heart attack. The manner of his death seemed somehow appropriate: In death, as in life, it was his own driving ambition to get to the top that ultimately brought him down.

———

Cameron, D. E. (1956). "Psychic Driving." *The American Journal of Psychiatry* 112 (7): 502–9.

The White Bear

It begins with a simple request. Sit in a room and say whatever comes into your mind. Do this for five minutes. You happily begin talking, and before you know it the time is up. But then the experiment takes an unexpected turn. The researcher says: "Please verbalize your thoughts as you did before, with one exception. This time, try not to think of a white bear. Every time you say 'white bear' or have 'white bear' come to mind, though, please ring the bell on the table before you."

Huh? A white bear? This shouldn't be too difficult. How often do you think about white bears anyway? But complying with the request proves tougher than anticipated. You start talking, and suddenly a white bear lumbers into your thoughts. Try as you might, you can't suppress the image. You try to distract yourself. You think of other things—an upcoming dentist appointment or a pattern on the floor. But nothing works. The bear keeps pushing its way back into your head.

In 1987 psychology professor Daniel Wegner devised the white-bear experiment and tested it on ten Trinity University undergraduates. He found that not one of them could stop thinking about white bears. Consider the frustration of one of his subjects (note: asterisks indicate bell rings signifying thoughts of a white bear):

> Of course now the only thing I'm going to think about is a white bear . . . Ummm, what was I thinking of before? See, if I think about flowers a lot * . . . I'll think about a white bear, it's impossible. * I could ring this bell over and over * and over * and over * and . . . a white bear * . . . and okay . . . And I'm trying to think of a million things to make me think about everything * but a white bear and I keep thinking of it over * and over * and over * and over. So . . . umm, hey, look at this brown wall. It's like every time I try and not think about a white bear, I'm still thinking about one, and I'm tired of ringing the bell.

But the experiment wasn't over. Wegner next asked the undergraduates again to say whatever came to mind, but now he gave them permission to think about white bears. Surprisingly, they didn't just think about the bears; they obsessed over them. They rang the bell like Quasimodo on steroids. By contrast, a control group of ten undergraduates who had received instructions to contemplate white bears from the outset rang the bell far less often. Wegner realized he had discovered a rebound effect—initial suppression of white bears leads to subsequent obsession about them.

So did all this have a point, or was Wegner just peculiarly interested in white bears? Yes, it did have a point. And no, Wegner was not a white-bear fanatic. The white-bear image was arbitrary. He borrowed it from an anecdote told by Dostoyevsky (an anecdote that, he sheepishly admitted, he read in *Playboy* magazine while a college student).

The point was that white bears represent unwanted thoughts. If you've ever found yourself unable to stop thinking about something—such as food, cigarettes, alcohol, or an ex—you're familiar with unwanted thoughts. Wegner's experiment revealed that the more we try to control what we think, the more control slips out of our grasp.

The real killer, of course, is the rebound effect. Perhaps your diet has been working. You've successfully fought off your cravings for weeks. You're not thinking about food. Then someone hands you a cupcake at an office party, and it's like a dam bursting. The next thing you know, you've downed the entire tray and are halfway through the bowl of nachos. Your suppression has turned into obsession.

And it gets worse, because the rebound effect can lead to an escalating cycle of suppression and obsession. Wegner writes, "The person becomes alarmed, noticing that an unusual degree of preoccupation is underway. This might produce a newly energized attempt at suppression, only to restart the cycle . . . Eventually, pathological levels of obsessive concern could result."

At which point you find yourself sitting in a corner, foaming at the mouth, madly screaming, "White bear! White bear! White bear!"

Wegner's experiment, as quirky as it is, is today considered a classic of modern psychology. One of the fun things about it is that you can try it yourself. Just put down this book and don't think about a white bear for a while. But don't say you haven't been warned. White bears, once not invited in, can be devilishly hard to get rid of.

Wegner, D. M., & D. J. Schneider (1987). "Paradoxical Effects of Thought Suppression." *Journal of Personality and Social Psychology* 53 (1): 5–13.

Lost in the Mall

Fourteen-year-old Chris sits at a table. A redheaded researcher in her late forties sits opposite him. She leans forward and looks him in the eye. "Chris, tell me what you remember about being lost in the mall when you were five."

Chris furrows his brow. "I remember that it was 1981 or 1982," he says slowly, as if struggling to recall. "We had gone shopping at the University City shopping mall in Spokane."

Then his voice quickens: "I was with the guys for a second, and I think I went over to look at the toy store, the Kay-Bee toys, and, uh, we got lost, and I was looking around and I thought, 'Uh-oh. I'm in trouble now.' You know. And then I . . . I thought I was never going to see my family again. I was really scared, you know?"

Chris's tone is confident. It is clear he remembers this frightening event. What he doesn't realize is that it never happened.

The researcher in this scene was Elizabeth Loftus, a professor of psychology at the University of California, Irvine. She had implanted a memory in Chris's head. Not with surgery, but through the power of suggestion. It was a trick she had performed often. In her book *The Myth of Repressed Memory* she boasted:

> I've molded people's memories, prompting them to recall nonexistent broken glass and tape recorders; to think of a clean-shaven man as having a mustache, of straight hair as curly, of stop signs as yield signs, of hammers as screwdrivers . . . I've even been able to implant false memories in people's minds, making them believe in characters who never existed and events that never happened.

Chris was a participant in Loftus's most famous experiment—the lost-in-a-mall study, conducted in the early 1990s. Subjects believed they were taking part in a study of childhood memories. They each received a booklet containing four short accounts, written by a relative, of events from their past. If they didn't remember an event, they were asked to write "I do not remember this." But if they did remember it, they were asked to elaborate both in the booklet and in a series of follow-up interviews.

What participants didn't know was that one of the narratives, about the subject getting lost in a shopping mall at the age of five, was fictitious. Their relatives, who were collaborating with the experimenter, provided enough personal details to make the story plausible.

The mere suggestion of the story was enough to implant a corresponding memory in the minds of many participants. Over a quarter of them (seven out of twenty-four) claimed to remember the event clearly. Many described it as an extremely vivid memory, and during follow-up interviews they freely supplied new details. When told that, in reality, they had never been lost in a shopping mall, they were dumbstruck. They insisted it must have happened. After all, they could remember it. Not until their relatives confirmed that it hadn't happened would they accept being mistaken.

Loftus rejects the videotape-recorder model of memory popularized by Wilder Penfield, in which our brains neatly file away everything we experience. She prefers to think of the mind as a bowl of water. "Imagine each memory as a teaspoon of milk stirred into the water," she writes. She describes all the memories in the mind constantly getting mixed around, blending and merging together into a cloudy, convoluted mess.

Which has to make you wonder about those childhood memories we all carry around—our first day at school, opening a present on our birthday, getting lost at a shopping mall. Did any of it really happen the way we remember? Or is it all just a product of our overactive imagination . . . or someone else's?

Loftus, E. F., J. A. Coan, & J. E. Pickrell (1996). "Manufacturing False Memories Using Bits of Reality," in Reder, L. M. (ed.), *Implicit Memory and Metacognition*: 195–220. Mahwah, NJ: Lawrence Erlbaum Associates.

CHAPTER FOUR

Bedtime Stories

A woman plays solitaire, waiting for her husband to return home. As she shuffles the cards, she feels the numbing fingers of sleep creep over her. Her eyelids grow heavy. "Just a little nap," she thinks as she leans forward to rest her head on the table. Within seconds, sleep overcomes her. Immediately, unbeknownst to her, her body begins a complex physical process. First, her body temperature and blood pressure drop. Her breathing grows shallow. Her heart rate slows, and her muscles relax. If someone were measuring her brain waves, they would see a gradual shift toward a slow, rolling pattern. If she remains asleep long enough—perhaps sixty to ninety minutes—her vital signs will abruptly change again. Her brain will become intensely active—as much so as if she were awake, as bizarre, irrational dreams flit through her head. Nearly all her muscles will lose their tone, held still by a temporary paralysis. She will have entered the phase of sleep known as rapid-eye-movement (REM) sleep—so named because of the characteristic movement of the eyes, back and forth, and up and down, beneath the eyelids. This stage of sleep could also be called nocturnal-erection sleep, because the sexual organs of both genders (and people of all ages) become engorged with blood during the twenty or thirty minutes this stage lasts, as though the body is making sure the plumbing still works. Throughout the remainder of the night, her body will cycle back and forth between REM and non-REM sleep. She'll awake unaware that any of this has happened.

Sleep and its physiological changes are mysterious, and for the most part unnoticed, phenomena. As our consciousness slips away, it can seem as though a phantom force takes control of our body. Faced with such an enigma, researchers have often resorted to peculiar analytical methods—as the experiments in this chapter demonstrate—to unlock sleep's secrets.

Sleep Learning

"My fingernails taste terribly bitter. My fingernails taste terribly bitter." A disembodied voice repeats the phrase over and over. A boy opens his eyes and lies very still in the darkness, listening to the words. He looks to his left and right. None of his camp mates seem to hear what he's hearing. They are all asleep in their cots. "My fingernails taste terribly bitter," the voice says again. The boy wonders if the voice is coming from inside his head. Is he going mad?

The boy was not going mad. Unbeknownst to him, he was a participant in a sleep-learning experiment devised by Professor Lawrence LeShan of William and Mary College.

In 1942 LeShan played a phonograph recording of the phrase "My fingernails taste terribly bitter" in a room where twenty young boys were sleeping at an Upstate New York summer camp. He played it in the middle of the night, after he felt sure none of them were awake. Competing with the chirping of crickets, the phrase repeated in the darkness 300 times a night, fifty-four nights in a row. The boys heard it in their sleep 16,200 times before the summer was over.

LeShan wanted to find out whether verbal suggestions given during sleep could influence waking behavior. All the boys bit their nails. So would repeated nocturnal exposure to a negative suggestion about nail biting cause them to abandon this nervous habit?

One month into the experiment, a nurse surreptitiously checked their nails during a routine medical examination. One boy seemed to have kicked the habit. LeShan boasted that skin of a healthy texture had replaced the "coarse wrinkled skin of the habitual biter."

But a week later, disaster struck. The phonograph broke. Eager not to abandon the experiment, LeShan improvised by delivering the suggestion himself, three hundred times a night. If any of the boys had wondered before what was going on, they now would have been completely bewildered as they woke to the sight of a grown man standing in the darkness insisting that his fingernails tasted terribly bitter.

Surprisingly, direct delivery of the suggestion had a greater effect. Within two weeks, seven more boys had healthy-looking nails. By contrast, a control group of twenty boys not exposed to the suggestion continued to bite away.

Why the sudden success at the end of the experiment? LeShan speculated that it was because his voice was clearer than the phonograph. Another theory would be that his midnight confessions thoroughly spooked the children. *If I stop biting my nails,* they probably thought, *the strange man will go away.*

LeShan achieved a 40 percent success rate. Does this mean

sleep learning works? For a long time, many researchers were inclined to believe so—especially since a string of other studies seemed to confirm the theory. For instance, during World War I a U.S. naval researcher reported success teaching sixteen cadets Morse code as they slept, though he never published his results. A 1947 study at the University of North Carolina found a group of students could learn a list of words faster if aided by a sleep-learning machine. A 1952 George Washington University study reported sleep instruction accelerated memorization of a list of Chinese words. There was also a widely repeated anecdotal account, disseminated by a tape-recorder salesman, of a housewife surreptitiously using the technique to train her husband to like salad.

By the late 1940s public interest in sleep learning was at an all-time high—fueled by spectacles such as a public demonstration of sleep learning, sponsored by a company selling learn-a-foreign-language phonograph records, in a storefront on Connecticut Avenue in Washington, D.C. Curious pedestrians stopped to watch as 1949's Miss Washington, Mary Jane Hayes, wearing a strapless bathing suit, climbed into a bed and pretended to doze as a machine whispered French phrases in her ear. "*Bon soir* . . . Good night . . . *Bon* . . . good . . . *le soir* . . . the night." One reporter covering the event joked, "Frankly, I'd rather spend my nights thinking about Miss Washington than about a French noun." Ms. Hayes, after changing her first name to Allison, later became a prominent figure in the dreams and fantasies of many young men when she played the title character in *Attack of the 50-ft Woman*.

An inventor named Max Sherover announced plans to market a commercial sleep-learning machine, which he called the Cerebrograph. It was a combination record player, clock, and pillow microphone. He secured testimonials from celebrities such as opera star Ramón Vinay, who claimed the device helped him memorize his lines. However, the gadget never caught on with the public, even when Sherover relaunched it with a new name—the Dormiphone.

In 1956 the scientific tide began to turn against sleep learning when William Emmons and Charles Simon published the results of a carefully controlled study conducted at Santa Monica College. The two researchers used an electro-encephalograph (an instrument that measures brain activity) to make sure their subjects were fully asleep—a precaution previous researchers had never taken—before reading them a list of nouns. Under these conditions, the sleep-learning effect disappeared.

Since that time, scientific interest in sleep learning has gone through ups and downs—though mostly downs. Much of the current research into the subject is conducted by high school students for science fairs. However, some informal studies carried out by Bill Steed of Emeryville, California, during the 1970s are worth mentioning. Steed chose frogs as his subjects, and motivational messages such as "Think positively" and "Don't let your past destroy your future" as their sleep lessons. (They must have been English-speaking frogs.) These same frogs went on to become regular champions at the Calaveras County frog-jumping competition (made famous by Mark Twain). So maybe there is something to the theory of sleep learning. After all, it's hard to argue with a high-jumping frog.

――――
LeShan, L. (1942). "The Breaking of a Habit by Suggestion during Sleep." *Journal of Abnormal and Social Psychology* 37: 406–8.

Eleven Days Awake

On the first day, Randy Gardner woke at six a.m. feeling alert and ready to go. By day two he had begun to drag, experiencing a fuzzy-headed lack of focus. When handed a series of objects, he struggled to recognize them by touch alone. The third day he became uncharacteristically moody, snapping at his friends. He had trouble repeating common tongue twisters such as *Peter Piper picked a peck of pickled peppers.* By the fourth day, the sand-clawed demons of sleep were scraping at the backs of his eyeballs. He suddenly and inexplicably hallucinated that he was Paul Lowe, a large black football player for the San Diego Chargers. Gardner, in reality, was white, seventeen years old, and 130 pounds soaking wet.

Gardner, a San Diego high school student, was the subject of a self-imposed sleep-deprivation experiment. He had resolved to find out what would happen to his mind and body if he stayed awake from December 28, 1963, to January 8, 1964, a total of 264 hours—eleven days. Assisting him were two classmates, Bruce McAllister and Joe Marciano Jr. They kept him awake and tracked his condition by administering a series of tests. They planned to enter the results in the Greater San Diego High School Science Fair. But transforming the ordeal from a science fair stunt into one of the most widely cited sleep-deprivation experiments ever conducted was the arrival of Stanford researcher William C. Dement, who flew down from Palo Alto to be with Randy as soon as he heard what was going on.

No one knew what Randy might experience as more days passed, or whether he might cause himself permanent brain damage, because only a handful of sleep-deprivation trials had ever been conducted. One of the earliest studies in this field had come to an inauspicious conclusion. In 1894 Russian physician Marie de Manaceine kept four puppies awake almost five days, at which point the puppies died. She reported that the research was "excessively painful," not only for the puppies but for herself as well. Apparently monitoring sleepy puppies 24/7 is hard work.

However, the few studies conducted on humans offered more hope. In 1896 doctors J. Allen Gilbert and George Patrick kept an assistant professor and two instructors awake in their lab at the University of Iowa for ninety hours. After the second night, the assistant professor hallucinated that "the floor was covered with a greasy-looking, molecular layer of rapidly moving or oscillating particles." But no long-term side effects were observed. Then, in 1959, two disc jockeys separately staged wake-a-thons to raise money for medical research. Peter Tripp of New York stayed awake for 201 hours while broadcasting from a glass booth in Times Square. Tom Rounds of Honolulu upped the ante by remaining awake 260 hours. Both Tripp and Rounds suffered hallucinations and episodes of paranoia, but after a few good nights' sleep they seemed fully recovered. It was Rounds's record Gardner hoped to beat, which is why he set his goal at 264 hours.

Meanwhile, Gardner valiantly pressed onward, struggling to stay awake. Nights were the hardest. If he lay down for a second, he was out like a light. So his high school friends and Dr. Dement kept him active by cruising in the car, taking trips down to the donut shop, blasting music, and playing marathon games of basketball and pinball. Whenever Gardner went to the bathroom, they made him talk through the door to confirm he wasn't dozing off. The one thing they didn't do was give him any drugs. Not even caffeine.

As more days passed, Gardner's speech began to slur, he had trouble focusing his eyes, he frequently grew dizzy, he had trouble remembering what he said from one minute to the next, and he was plagued by more hallucinations. One time he saw a wall dissolve in front of him and become a vision of a forest path.

To make sure he wasn't causing himself brain damage or otherwise injuring his health, his parents insisted he get regular checkups at the naval hospital in Balboa Park—the family's health-care provider since his father served in the military. The doctors at the hospital found nothing physically wrong with him, though he did sporadically appear confused and disoriented.

Finally, at two a.m. on January 8, Gardner broke Rounds's record. A small crowd of doctors, parents, and classmates gathered to celebrate the event. They cheered wildly, and Gardner, busy taking calls from newsmen, responded with a V-for-victory sign. Four hours later, he was whisked away to the naval hospital where, after receiving a brief neurological checkup, he fell into a deep sleep. He woke fourteen hours and forty minutes later, feeling alert and refreshed.

Gardner's world record didn't last long. A mere two weeks later, papers reported that Jim Thomas, a student at Fresno State College, managed to stay awake 266.5 hours. The *Guinness Book of Records* subsequently recorded that in April 1977 Maureen Weston, of Peterborough, Cambridgeshire, went 449 hours without sleep while participating in a rocking chair marathon. However, Gardner's feat remains the most well-remembered sleep-deprivation trial. To this day, no one knows the maximum amount of time a human can stay awake.

As of 2007, Gardner remains alive and well, having suffered no long-term ill effects from his experience. Despite sleep deprivation being the source of his fifteen minutes of fame, he insists he's really not the type to pull all-nighters and says he's maintained a sensible sleep schedule since his youthful stunt. He does admit to lying awake some nights, but attributes this to age, not a desire to beat his old record.

Ross, J. (1965). "Neurological Findings After Prolonged Sleep Deprivation." *Archives of Neurology* 12: 399–403.

Shaken, Not Stirred

It's three in the morning and you're trying to get to sleep. But you're not having much luck because you're stuck in a cramped seat on an airplane cruising at thirty thousand feet. Turbulence keeps shaking you. Lights in the cabin flash on and off. People wander up and down the aisle. Somewhere a baby is screaming. How in the world, you wonder, are you supposed to get any rest?

If, in the future, you find yourself in this situation, you might want to reflect on an experiment conducted in 1960 by Ian Oswald, a professor at Edinburgh University. In his lab, he asked subjects to try to fall asleep while being exposed to far more intrusive stimuli than you would experience on a typical plane ride—even given the ever-worsening conditions of economy class. The title of his study hints at the bizarre setting he placed his volunteers in: "Falling asleep open-eyed during intense rhythmic stimulation."

Three young men in their early twenties served as Oswald's guinea pigs. Testing them one at a time, he asked each of the subjects to lie down on a couch. He carefully attached one end of a piece of tape to each eyelid and the other end to the subject's forehead, keeping his eyes pried open. Steam from a boiling kettle in the room prevented the test subject's eyes from drying out. Next, Oswald placed electrodes on the subject's left leg. The electrodes produced a painful shock that caused the foot to bend sharply inward involuntarily. Oswald programmed the shocks to occur in a regular, rhythmic pattern. He also positioned a bank of bright flashing lights two feet in front of the man's face. With his eyes taped open, he couldn't avoid looking at these lights. Finally, Oswald turned on some blues music. The music, he noted laconically, "was always very loud."

Having placed each of his three subjects in this unfortunate situation—music blaring, eyes pried open and staring at flashing lights, foot jerking rhythmically from electric shocks—Oswald sat in a corner of the room and waited for them to do something that would seem unlikely in such a circumstance: fall asleep.

One subject was sleep deprived going into the test, having only had one hour of sleep the night before. The other two subjects were fully refreshed and awake. However, it turned out not to make any difference. Within eight to twelve minutes, all three men were asleep. At least, they showed all the signs of being asleep. Their heartbeat slowed, their pupils constricted, and their brain waves, measured by an EEG, displayed a low-voltage slow-wave pattern characteristic of sleep. In addition, the subjects reported afterward feeling as though they had fallen asleep.

Acknowledging possible skepticism of the claim that his subjects fell asleep, Oswald phrased his words carefully:

> It seems reasonable to believe that each of these volunteer subjects did go to sleep, but it will be remembered that there is no clear dividing line between wakefulness and sleep, and it is no part of my present concern to insist that subjects crossed any such dividing line, only to claim that there was a considerable fall of cerebral vigilance, and a large decline in the presumptive ascending facilitation from the brain-stem reticular formation to the cerebral cortex.

If your boss ever catches you napping at your desk, Oswald's wording could offer a convenient excuse: "No, I wasn't sleeping. I was merely experiencing a large decline in the presumptive ascending facilitation from the brain-stem reticular formation to the cerebral cortex."

Oswald performed a second test, in which he seated two new subjects each in a chair. Again, he taped their eyes open, played loud blues music, and flashed lights in their eyes. But instead of shocking their legs, he asked them to bang their elbows up and down and tap both feet in time with the music. Required to keep moving, these subjects did not drift into an extended period of sleep as the men in the first study had. However, Oswald did observe them repeatedly drifting off into spells of sleep that lasted from three to twenty seconds. During these microsleeps, their brain waves slowed and they stopped moving their limbs. Then they would come to with a start and begin moving again.

In one of his subjects, Oswald observed fifty-two of these pauses within twenty-five minutes. However, the pauses apparently happened without the subject's awareness, because the young man later emphatically maintained he had only paused once.

Oswald's results seem hard to believe. How could someone fall asleep under such conditions? Oswald explained it as a peculiar response of the brain to extremely monotonous sensory stimulation. Instead of becoming aroused by the stimulation, the brain becomes habituated to it and shuts down. He likened it to the trance effect tribal dancing induces. You may have experienced the effect yourself if you've driven down a highway for an extended period. It may be the middle of the day and you may have the radio blasting, but the road just keeps rolling along, and your mind wanders off. Moments later you come to with a start, aware that you have zoned out. You may not think you were actually asleep, but from a practical point of view there isn't much difference. As Oswald would put it, the presumptive ascending facilitation from your brain-stem reticular formation to your cerebral cortex was momentarily in decline.

So, to return to the airplane scenario, it's not the noise and lights, per se, that prevent you from falling asleep. It's the fact that they're not monotonously rhythmic. Airlines could remedy this situation by installing vibrating seats, pulsing lights, and continuously looping baby screams. Passengers would soon be drifting off into dreamland, whether they wanted to or not. Electric shocks would, of course, be reserved for business class.

———

Oswald, I. (May 14, 1960). "Falling Asleep Open-Eyed During Intense Rhythmic Stimulation." *British Medical Journal* 1: 1450–55.

Let Sleeping Cats Hunt

The cat freezes in place. It has seen its prey. Slowly it moves forward,
sliding its forelimbs across the floor. It freezes again. And then—
pounces. A vase crashes t the floor. A light switches on. "What
was that? What's going on?" a voice cries out. "Oh, it's nothing,"
another voice says. "It's just the cat sleepwalking again."

Do cats sleepwalk? Veterinarians report that children often
ask them this question, and it does seem like a natural topic
to be curious about. After all, humans sleepwalk. Why
shouldn't cats? The simple answer is, no, cats do not sleepwalk. However, occasionally, under certain circumstances,
they can do something like it.

In 1965 a French neurophysiology researcher named
Michel Jouvet was trying to pinpoint the parts of the brain
responsible for inducing sleep. His investigative procedure
consisted of damaging different parts of cats' brains and
noting what effect this had on the cats' sleep. He had already
learned that when he damaged a cluster of cells called the
nuclei of the raphe, located in the brain stem, cats would
barely sleep at all. They became insomniacs, shuffling around,
unable to settle into their customary catnaps. This led to his
conclusion that the nuclei of the raphe, which secrete the
chemical serotonin, tell the brain to go to sleep.

His next experimental target was a part of the brain stem
called the locus coeruleus. He operated on thirty-five cats,
creating a lesion on this part of each of their brains.

Sleep, in a normal cat, follows a predictable pattern. First
the cat will settle into a period of light sleep, during which it

often curls up into a ball. Its brain waves exhibit a slow-wave pattern, but its muscles remain slightly tense. After about twenty minutes of this, the cat progresses into the dreaming stage of sleep, known as rapid-eye-movement (REM) or paradoxical sleep (PS)—"paradoxical" because the brain waves during this period are paradoxically as energetic as they are during wakefulness. During PS, the cat's muscles go completely limp, except for occasional brief twitching motions in the paws, tail, and ears. If you own a cat, you've probably seen it twitch in this way, as though dreaming of chasing mice, or some other object of cat interest.

As Jouvet's cats recovered from the surgery, they began to display strange sleep behavior. They would fall asleep normally enough, and nap through twenty minutes of light sleep, but things got weird when they proceeded into paradoxical sleep. Many of them would abruptly lift their heads and look around. They might even stand up. All the while, they appeared to be fast asleep—except that they were fully mobile.

Jouvet checked that the cats were really asleep. Although their eyes were open, their pupils were constricted and their nictitating membranes (the white membranes inside their eyes) were relaxed, as is typical during sleep. They didn't respond to bright lights, nor react to pinching. In addition, their brain waves showed a pattern characteristic of dream activity.

For the next few minutes, the actions of these somnambulistic cats grew progressively more violent and erratic. They stalked around and showed signs of rage. They even leaped and pounced on nonexistent objects. The cats were, in essence, acting out their dreams, stalking imaginary prey. Jouvet reported that the behavior of the cats could "be so fierce as to make the experimenter recoil."

If the cats pounced violently enough, they would jolt themselves awake. At which point, they would shake their heads sleepily, as if to say, *Where am I? What's happening?*

Jouvet eventually concluded that the locus coeruleus must be responsible for sending a signal to muscles telling them to remain paralyzed as the cat dreams. By making a lesion on this region, he had disrupted the signal. He called the phenomenon "paradoxical sleep without atonia" (*atonia* meaning paralysis, or lack of muscle tone).

The story of cats seemingly acting out their dreams doesn't end there. The phenomenon turned out to be more complicated than Jouvet thought. (Anything having to do with brain research usually is more complicated than first appearances suggest.)

Other researchers were skeptical of Jouvet's results. So during the 1970s Adrian Morrison of the University of Pennsylvania replicated Jouvet's experiment. He succeeded in producing cats that stalked imaginary objects in their sleep, or savaged the towels on which they lay, thus confirming Jouvet's basic claim. However, he disputed Jouvet's interpretation that the cats were acting out their dreams. He discovered he could engineer a variety of eccentric sleep behaviors on demand, depending on the location of the lesion he made.

By making a relatively small lesion below the locus coeruleus, Morrison produced cats that simply lifted their heads and looked around during PS. A larger lesion caused predatory attack behavior. Still another type of lesion triggered behavior resembling sleepwalking. These cats would stand up and march forward in a straight line. They never deviated from their path, not even to sniff at an anesthetized rat placed directly in front of them. They simply stumbled over it and continued onward until they ran into an immovable obstacle such as a wall.

Morrison also observed that the cats never exhibited certain types of behavior in their sleep. They never ate, drank, or engaged in sexual activity. He concluded that the lesions in the brain stem were not releasing *general* dream-related activity. Instead, they were causing the cats to act out *specific* kinds of behavior, such as aggression and locomotion.

It should be noted that the behavior of these cats was not analogous to sleepwalking in humans. People who sleepwalk typically do so during light sleep, not during the deeper, dreaming state of PS. However, there are cases of people who move violently and even run during PS. They're suffering from REM Sleep Behavior Disorder, a condition discovered by clinicians Mark Mahowald, Carlos Schenck, and colleagues at the University of Minnesota, thanks to their knowledge of the experiments by Jouvet and Morrison. As with the cats, sufferers often go into attack mode during these episodes, which can be extremely dangerous both for them and for their partners. One man woke to find he was strangling his wife. He had been dreaming of killing a deer. (And you think you have it rough because your partner snores!)

Meanwhile, if you wake to find your carpet shredded and your vases broken, don't believe your cat when it plays innocent and offers the sleepwalking excuse. The odds are it knew perfectly well what it was doing. It's just messing with you, as cats often do.

Jouvet, M., & F. Delorme (1965). "Locus coeruleus et sommeil paradoxal." *Comptes rendus des séances de la Société de Biologie* 159 (4): 895–99.

What Dreams May Come

The reels of the movie projector turn, and a grainy black-and-white image springs to life on the screen. A man sits in the darkened laboratory, watching the picture. A researcher standing behind the projector addresses him, "What you are going to see is footage shot by the anthropologist Géza Róheim. It shows a subincision initiation ritual practiced by Australian Aborigines. Please pay careful attention. We ask that you not look away, no matter how much you might feel like doing so." The man nods that he understands.

In the movie, a group of naked aboriginal men—four older and one younger—are standing around. The men crouch down on all fours, and the young man lies face upward on their backs. Other men appear and hold the youth down. A medicine man walks on screen. He is holding a sharp stone. Abruptly he grasps the young man's penis and moves the stone toward it.

The man watching this squirms in his seat. "Tell me he's not going to . . . ," he mutters. "Oh no! Oh my God! Yes, he is! I can't watch!"

"Please continue to watch," the experimenter immediately instructs.

The medicine man uses the stone as a knife, deftly slicing the underside of the youth's penis open lengthwise, from the tip toward the base.

"Oh, that's got to be painful," the viewer moans. And indeed, the youth's face is contorted in agony. He is being held down tightly. The view changes, and now the bleeding penis is being held over a fire, cauterizing it.

"Oh, good grief!" the viewer exclaims. He twists in his chair,

holding his head at an angle as though it's physically painful for him to continue watching.

Finally, the penis-burning scene ends, and the action switches to a hairdressing ritual. The film closes with Aborigines performing a rhythmic dance.

"Thank you for watching," the experimenter says as he turns off the projector. "Now please prepare yourself for sleep."

Do emotionally charged events that happen to us while we are awake influence the content of our dreams? If they do, in what way? These were the questions posed by psychologists Herman Witkin and Helen Lewis. To find the answer, they conducted an experiment in 1965 at the State University of New York Downstate Medical Center.

The methodology of the experiment seemed straightforward. The researchers exposed subjects, right before they went to bed, to a dramatic event—something that would evoke a strong psychological reaction. After falling asleep, the subjects were periodically awakened and asked what they were dreaming. Witkin and Lewis hoped the arousing event would act as a "tracer element." In theory, its influence on dream content would be readily identifiable because the event itself was so unmistakable, and its transformations through the various stages of sleep could be followed.

The subjects were men from various blue-collar professions—post-office employees, guards, airplane-factory workers, telephone engineers, and bakers—who worked at night and slept during the day. They each received ten dollars a day for participating.

The emotionally arousing events the researchers exposed them to were three different films—and they definitely weren't Hollywood tearjerkers. Forget *Doctor Zhivago* or *Gone with the Wind.* The researchers wanted in-your-face shock value, something the men wouldn't be able to ignore. One film showed the subincision initiation ritual as described above. A second film showed an obstetrician delivering a child with the help of a Malmström Vacuum Extractor. The researchers wrote that the movie showed

> the exposed vagina and thighs of the woman, painted with iodine, in other words, brown. The arm of the obstetrician is seen inserting the vacuum extractor into the vagina; the gloved hands and arms of the obstetrician, covered with blood, are then shown pulling periodically on a chain protruding from the vagina. The cutting motion of an episiotomy is also shown. The baby is then delivered with a gush of blood. The film ends by illustrating that only a harmless swelling of the skin of the baby's head results from the vacuum extraction method.

The third film was perhaps the most horrific of all. It was shot at the State University of New York's primate laboratory:

> In the film a mother monkey is shown eating her dead infant. The mother is seen hauling her dead baby about, dragging it with her by arms and legs, and nibbling at it. One scene, toward the end, shows her eating the lips and protruding tongue of the baby.

A fourth film was also shown. However, it was an unexciting piece of footage, an educational travelogue about the western United States, intended to provide the researchers with data about the dream response to neutral content.

Subjects watched one film per session, and then immediately went to bed. To gain as much information as possible about the train of thought between seeing the film and falling asleep, the researchers asked the men to speak aloud whatever ideas popped into their heads until sleep overcame them. To facilitate this, earphones fed white noise into the men's ears, and halves of Ping-Pong balls, over which shone a diffuse red light, covered their eyes. The researchers reported that, thanks to this technique, the men were able to continue talking almost right up to the instant they fell asleep. In addition to being awakened at fixed intervals and asked to report their dreams, the men participated in postsleep interviews.

So did the presleep experiences influence the men's dreams? Witkin and Lewis declared unambiguously that they did. "It is quite evident that elements from the exciting presleep stimuli we used often appeared in the subsequent dreams in transformed fashion," they reported. The key phrase in this statement is "in transformed fashion." In no case did elements from a film appear directly in a dream. People who watched the birth film didn't dream of delivery room scenes. Nor did the men exposed to the subincision ritual dream of dancing Aborigines. Instead, elements from the films manifested themselves in the dreams in symbolic form. At least, that's what the experimenters said happened. Whether you agree with this conclusion depends a great deal on the faith you place in Freudian psychoanalysis.

For instance, after viewing the birth film, one subject dreamed he was in a hot closet taking a piece of cake out of a brown paper bag. Witkin and Lewis viewed this as an obvious symbolic reference to the movie:

> The female body seems symbolized as a hot closet (amended in the inquiry into a hot moist closet in danger of spontaneous combustion) in which there is a "common grocery type bag"—the vagina.

The birth film evoked in another subject a dream of "sort of a troop-carrier plane, people, parachutists, jumping out of the airplane." Again, Witkin and Lewis provided a translation:

> The delivery from the womb is represented by congruently structured and functioning mechanical objects: the airplane with its regularly opening door ejecting the parachutists. The periodic motion of the obstetrician's hands pulling on the chain is represented in the periodic opening of the airplane door.

Similar interpretations were applied to dreams following the other movies. A dream of two cowboys, one holding a gun on the other, that occurred after the subincision film was thus a "classic symbolic representation of the penis as a gun." The monkey film yielded a reverie about a blue green frog sitting in a pool of water. This initially stumped the researchers, until they learned, on deeper inquiry, that as a child the test subject used to torture frogs:

> He would throw them across a brick-wall incinerator and kill them. . . . The "frog" in the hypnagogic interval was thus connected with memories of his own cruelty.

The neutral travelogue film, as the experimenters had predicted, did not yield any dreams containing readily identifiable Freudian imagery.

While the connection between these dream images and the presleep stimulus was apparent to the researchers, it was not so to the subjects themselves. The researchers reported that "our subjects were sometimes quite vehement in denying any connection between the presleep event and their dreams." Evidently the subjects were not sufficiently indoctrinated into Freudianism. The subjects persisted in their "denial" (as the experimenters labeled it) even after shown how obvious the connections were. Witkin and Lewis tried to explain to one subject that the brown paper bag in his dream had to be a reference to the brown-iodine-stained vagina in the birth film:

> Yet, even with the occurrence of the element "brown" in his imagery of the brown paper bag with cake in it in a hot closet . . . he was not able to see the connection between these dreams and the content of the film he had seen.

One can sense the rising frustration of the researchers, as though they wanted to shout out: *Why can't you guys see the connection? How blind can you be?* However, since it was their experiment, Witkin and Lewis got to have the last word.

The dream study was not a career highlight for either Witkin or Lewis, and they only made modest claims about it, offering their results as evidence that their procedure could be used to study the ways in which "thoughts and feelings stirred in the waking state are represented in subsequent dreams." Witkin is far better remembered for designing tilting rooms for the U.S. Air Force, which allowed him to investigate how people determine their orientation in topsy-turvy environments such as fighter jets. Lewis is best remembered for her studies of shame. Of the two of them, Lewis was the Freudian psychoanalyst—so you have to assume she exerted the major influence on the dream study.

Quite a few other researchers have studied the relationship between waking experiences and dreams. In a 1968 study, subjects wore red-filtered goggles continuously for five days, causing their dreams to take on a red hue. Even the British Cheese Board has contributed to this field. In 2005 it sponsored a "Cheese & Dreams" study, in an effort to disprove the old legend that cheese causes nightmares. For seven days two hundred volunteers ate twenty grams of cheese half an hour before going to bed. Varieties included Stilton, Cheddar, Brie, and Red Leicester. No one reported any nightmares, but many did have unusual dreams. One subject dreamed of finding celebrity chef Jamie Oliver cooking dinner in her kitchen. The Freudians would doubtless have something to say about that. Another dreamed of having a drunken conversation with a dog.

The authors of the cheese study suggested that different types of cheese may have varying effects on dreams. For instance, Cheddar seems to cause more dreams about celebrities, whereas Brie promotes nice, relaxing dreams. Which has to make you wonder, what kind of dreams might cutting the cheese as you lie in bed cause? Probably nightmares, at least for your sleeping partner.

Witkin, H. A., & H. B. Lewis (1965). "The Relation of Experimentally Induced Presleep Experiences to Dreams." *Journal of the American Psychoanalytic Association* 13 (4): 819–49.

Do Amnesiacs Dream of Electric Tetris?

You are dreaming. You see large rectangular blocks tumbling downward, rotating as they fall, landing on top of an ever-growing pile of blocks. Suddenly you're shaken awake. You have no idea where you are or who this person wearing a white lab coat is. "What were you dreaming?" the stranger asks. You answer, "Little squares coming down on a screen and trying to put them in place."

You, in this scenario, are a participant in an experiment conducted by Harvard researcher Robert Stickgold. In 2000

Stickgold arranged for twenty-seven people to play the video game Tetris every morning and night for three days. Ten of these people were experienced Tetris players, twelve were novices, and five were amnesiacs suffering from "extensive, bilateral medial temporal lobe damage." (Translation: They could barely remember events from one minute to the next.) Stickgold repeatedly awakened his subjects during their first hour of sleep and asked them to describe their thoughts. Almost two-thirds of the participants, including three of the five amnesiacs, reported dreams of falling Tetris cubes.

The surprise was that the amnesiacs would dream of Tetris. After all, their amnesia was so severe they couldn't remember playing the game, the experimenter, or that they were participating in an experiment. One amnesiac, after playing her nightly game of Tetris, expressed shock when she returned from taking a shower and found the experimenter sitting in her room. In the intervening minutes she had forgotten who he was. So why had the amnesiacs' sleeping minds held on to the Tetris memory?

Stickgold argues that his experiment reveals that our dreams do not draw on concrete memories, the kind that are stored in the hippocampus. If dreams did, the amnesiacs would not have dreamed of Tetris, since it was this area of their brains that was damaged. Instead, dreams tap into the vaguer, more abstract imagery stored deep within the cortex. Which explains why they often seem so illogical.

It has yet to be determined whether these results hold true for Pac-Man, Asteroids, Frogger, or Space Invaders.

Stickgold, R., A. Malia, D. Maguire, D. Roddenberry, & M. O'Connor (Oct. 13, 2000). "Replaying the Game: Hypnagogic Images in Normals and Amnesics." *Science* 290: 350–53.

CHAPTER FIVE

Animal Tales

London, March 1626. Sir Francis Bacon, the founder of modern science, is on his knees, burying a dead chicken in the snow. Carefully he packs the ice crystals around the carcass of the bird. It is slow work and he is poorly dressed. As he finishes up, he sneezes.

Bacon was not shivering out in the cold just for fun. He was doing it for the sake of science. At the age of sixty-five, he was conducting his first-ever experiment, attempting to find out whether snow could be used to preserve meat. Sadly, the elements got the upper hand, and thus it was also his last experiment. He caught pneumonia and died a few weeks later—felled by a frozen fowl.

The story of Sir Francis and the chicken is not merely of biographical interest. It highlights the central role animals have played throughout the history of experimental science. It is hard to imagine where modern science would be without the untold number of animals who have served as guinea pigs in experiments. Researchers sometimes study animals for the sake of increasing scientific knowledge about animal behavior, but more often they're using animals as a stand-in for humans. The idea is to try something on an animal first, and hope it works the same way in humans. Whatever the reason for the use of animals, it is a truism that, as we shall see in this chapter, the behavior of the researchers is often far more curious than the behavior of the animals they're studying.

Elephants on Acid

Tusko the elephant led a peaceful life at the Oklahoma City Zoo. There were his daily baths, playtime with his mate, Judy, and the constant crowds of people peering at him from the other side of the fence. Nothing out of the ordinary. So when he awoke in his barn on the morning of Friday, August 3, 1962, he could hardly have foreseen what that day held in store. He was about to become the first elephant ever given LSD. He would simultaneously become the recipient of the largest dose of that drug ever administered to any creature— a record that stands to this day. If he had known what was about to happen to him, he probably would have made a run for it.

The experiment was the brainchild of two doctors at the University of Oklahoma School of Medicine—Louis Jolyon "Jolly" West and Chester M. Pierce—as well as Warren Thomas, director of the Oklahoma City Zoo. All three men were impressed by the effects of LSD and were eager to learn more about the drug's pharmacological properties. So, in an effort to expand the frontiers of psychiatric knowledge, they turned their restless imaginations to elephants.

In fairness to the men, they were not alone in their interest in LSD. At the time, there was a great deal of research into the drug, for a number of reasons. Doctors were fascinated by LSD because of the powerful effect it had on patients. It seemed like a wonder drug, able to heighten patients' self-awareness, facilitate the recovery of memories, and entirely alter patterns of behavior. There were reports of it curing alcoholism almost overnight. Many hoped it might have a similar effect on schizophrenia. It wasn't until the mid-1960s that doctors became more concerned about the drug's dangers—and its popularity among counterculture youth—which prompted the U.S. government to ban its use.

More shadowy forces were also promoting LSD research. The Central Intelligence Agency was extremely curious about the military applications of the drug. Could it be used as a debilitating agent in chemical warfare or as a brainwashing tool? To get answers to these questions, the agency was funneling large sums of money to researchers throughout America. Virtually everyone who was anyone working in the behavioral sciences at the time received CIA money, though many weren't aware of it because the agency disseminated funds through various front organizations. There is no evidence, however, that the CIA played any role in the elephant experiment.

Finally, psychiatrists were interested in LSD because its effects seemed to mimic the symptoms of mental illness. It produced what they called a "model psychosis." Quite a few doctors took the drug to gain a more intimate sense of what their patients might be experiencing, and researchers were testing LSD on animals in the hope they could experimentally simulate mental illness and thus examine the phenomenon in a more controlled way. Giving LSD to an elephant was, in a sense, the logical outgrowth of such studies.

But the three researchers were particularly interested in elephants for additional reasons. First, the animal's large brain size offered a closer analog to a human brain. Second, male elephants experience periodic episodes of madness known as musth. When they go into musth, the males become highly aggressive and secrete a strange sticky fluid from their temporal glands, which are located between their eyes and ears. West, Pierce, and Thomas reasoned that if LSD truly did trigger temporary madness, then it might cause an elephant to go into musth. If this happened, it would be a powerful validation of LSD's ability to produce a model psychosis. Best of all, the onset of musth could be easily confirmed by looking for the secretion of the sticky fluid.

This was the scientific rationale offered for the experiment. But a small element of ghoulish inquisitiveness must also have been involved. After all, *what would an elephant on acid do?* It's hard not to be curious.

On the morning of August 3, the experimenters were ready. Thomas had arranged for the use of Tusko, a fourteen-year-old male Indian elephant. The pharmaceutical company Sandoz had provided the LSD. Just one thing was missing—knowledge of the appropriate amount of LSD to give Tusko. No one had given an elephant LSD before.

LSD is one of the most potent drugs known to medical science. A mere twenty-five micrograms—less than the weight of a grain of sand—can send a person tripping for half a day. But the researchers figured an elephant would need more than a person, perhaps a lot more, and they didn't want to risk giving too little. Thomas had worked with elephants in Africa and knew they could be extremely resistant to the effect of drugs. So they decided to err on the side of excess. They upped the dose to 297 milligrams, about three thousand times the level of a human dose. This, in hindsight, proved to be a mistake.

At eight a.m. Thomas fired a cartridge syringe into Tusko's rump. Tusko trumpeted loudly and began running around his pen. For a few minutes his restlessness increased, then he started to lose control of his movements. His mate, Judy, came over and tried to support him. But suddenly he trumpeted one last time and toppled over. His eyeballs rolled upward. He started twitching. His tongue turned blue. It looked like he was having a seizure.

The researchers realized something had gone wrong and took measures to counteract the LSD. They administered 2,800 milligrams of an antipsychotic, promazine hydro-chloride. It relieved the violence of the seizures, but not by much. Eighty minutes later, Tusko was still lying panting on the ground. Desperate to do something, the researchers injected a barbiturate, pentobarbital sodium, but it didn't help. A few minutes later, Tusko died.

If ever a situation justified an exclamation of "Oh, crap!" this was it. Tusko's death had definitely not been part of the plan. He was only supposed to go a little mad—run around his enclosure, trumpet a few times, maybe secrete some fluid to indicate he had gone into musth—and then a few hours later be right as rain. He shouldn't even have had a hangover. But instead, the researchers now had a dead elephant at their feet and a lot of explaining to do.

What had happened? Frantically, West, Pierce, and Thomas tried to figure that out. Had the LSD concentrated somewhere in Tusko's body, increasing its toxicity? Had they hit a vein with the cartridge syringe? Were elephants allergic to LSD? They really had no clue. An autopsy performed later determined Tusko died from asphyxiation—his throat muscles had swollen, preventing him from breathing. But why his throat muscles had done this, the researchers didn't know. In an article published a few months later in *Science*, they simply noted, "It appears that the elephant is highly sensitive to the effects of LSD."

Meanwhile, reporters had immediately learned of the experiment and were phoning the zoo, trying to find out details. Lurid headlines appeared in papers the next day— FATAL RESEARCH: DRUG KILLS ELEPHANT GUINEA PIG! and ELEPHANT DIES FROM NEW DRUG. The local paper, the *Daily Oklahoman*, ran the headline SHOT OF DRUG KILLS TUSKO on its front page. Beneath these words, a photo showed West bending over the lifeless body of the elephant.

Some of the most sensational details of Tusko's death first appeared in these news stories published the day after the experiment. However, much of the reporting resembled a game of Chinese whispers. What information came from West, Pierce, and Thomas, and what came from the imaginations of the reporters wasn't always clear. For instance, the Associated Press reported that the amount of LSD given to Tusko "was less powerful than the contents of an aspirin tablet." It's hard to imagine one of the experimenters saying this. What was probably said, if anything, was that the amount of drug given to Tusko was comparable, in weight, to the amount of medicine in an aspirin tablet—although the potency of the two drugs is vastly different.

The same AP story stated that "one of them, Dr. L. J. West, professor of psychiatry at the University of Oklahoma, had taken a dose of the drug Thursday." The story dropped this bombshell and then casually moved on, as if scientists self-dosing with LSD before experiments was standard operating procedure. The modern reader, coming across the statement, can't help but do a double take—*Huh? Come again?* If West took the drug Thursday he could still have been strung out on Friday when Tusko received the drug. This information, if true, would cast the experiment in a new, even more psychedelic-hued light.

The *Daily Oklahoman*, whose reporter presumably had better access to the experimenters, made a similar claim, but stated it more ambiguously: "Dr. West said he and Dr. Chester Pierce, chief of psychiatry at Veterans Administration Hospital, had taken LSD prior to Tusko's injection." This could mean they took the drug the day before, or a year before.

It's difficult to know the truth. The researchers filmed the entire experiment, but the film remains hidden away in an archive at UCLA, where West later went to work. It has never been made accessible to the public. However, those who have seen it report that all three researchers appear perfectly sober throughout the experiment. This suggests reporters once again misinterpreted something they had heard.

Whatever the case may be, the idea of West and Pierce, high on acid, stomping around an elephant pen, quickly took root and was often repeated in retellings of the story. The Church of Scientology, in its attacks on the psychiatric profession, later singled out West and hammered him with this accusation, portraying him in their publications as an out-of-control researcher who was "evidently still under [LSD's] influence at the time he sloshed through the beast's entrails, performing an 'autopsy' which he recorded on film." (This accusation was at least partially incorrect since the autopsy was not filmed.)

One more detail stands out from the news stories. The *Daily Oklahoman* quoted Dr. Thomas as saying that Tusko "was no toy. He was getting hard to manage and had to be handled strictly from a distance. He might have been a potential killer." To many this sounded like a callous attempt to legitimate Tusko's death. But Thomas wasn't finished. The experiment *shouldn't* be considered a failure, he opined. After all, they had learned that LSD is lethal to elephants. This was potentially useful information. "Maybe LSD would be a more effective way of destroying herds in countries where they are a problem," he suggested. One elephant on acid wasn't enough. Thomas was imagining entire herds of elephants, roaming the African savanna, tripping on LSD before stumbling forlornly to their deaths.

To the scientific community, the entire experiment was an embarrassment. Other researchers took West, Pierce, and Thomas to task for so clumsily misjudging the appropriate amount of LSD to give Tusko. Paul Harwood, a veterinary biologist, wrote a letter to *Science* describing what they had done as an "elephantine fallacy." But, oddly, the experiment made the researchers minor celebrities within the counter-culture. West later remarked that when he was studying hippies during the late 1960s, he could gain instant access to their community by identifying himself as the guy who had given LSD to an elephant.

But the final chapter in the elephants-on-acid story was not yet written. It still had to become "elephants"—plural!

In 1969 West joined the faculty of UCLA. Ronald Siegel, a professor of psychopharmacology who later became one of his colleagues there, became interested in the experiment. Questions about it lingered. For instance, was it the LSD that had killed Tusko, or was it the combination of drugs? And could LSD actually induce musth? In 1982 Siegel decided to find out. He later said, "I couldn't let the obvious mistakes and procedural errors of this 'experiment' remain uncorrected." He set out to give LSD to more elephants.

Going into the project, Siegel had a number of advantages over his predecessors. For a start, he was one of the world's leading experts on the effects of hallucinogens on animals. He also had his predecessors' example to learn from. Basically, he knew what not to do.

Siegel obtained access to two elephants (one male, one female) that lived in a barn at an undisclosed location. It is rumored he had to sign an agreement promising to replace the animals in the event of their deaths. Instead of using a cartridge syringe to deliver the drug, Siegel put the LSD in the animals' water—after first denying them water for twelve hours to make sure they would be thirsty. This ensured a more gradual entry of the drug into their systems. He tested the elephants at two different dosages—a low dose of .003 mg/kg and a high dose of .1 mg/kg. The high dose was equivalent, in terms of amount per body weight (but not in terms of the absolute amount of the drug), to what Tusko had received.

The elephants didn't topple over dead. That was the good news. So what did they do? Many animals exhibit extremely unusual behavior under the influence of LSD. Spiders spin highly regular webs, goats walk around in predictable geometric patterns, and cats adopt a kangaroo-style posture in which they sprawl their legs and extend their claws and tail. The elephants, however, didn't do anything so bizarre. At the low dose Siegel observed behavioral changes such as increased rocking and swaying, vocalizations such as squeaking and chirping, and head shaking. At the high dose, the elephants initially exhibited aggressive behavior before slowing down and becoming sluggish. The male gave himself an extended hay bath. Twenty-four hours later, both animals were back to normal.

Siegel's experiment cast doubt on the idea that it was the LSD that killed Tusko, but he could not rule out the possibility. After all, the amount given to Tusko may have exceeded some threshold of toxicity. Siegel also noted that LSD did not induce a musth-like state, which disproved West, Pierce, and Thomas's original hypothesis.

Siegel's experiment demonstrated what should have happened to Tusko in 1962. The sad reality, however, is that Tusko died. Ironically, his death ensured that, instead of ending up as a footnote in obscure articles about zoological pharmacology, Tusko gained a permanent place in pop culture. In addition to articles and books (such as the one you're reading now), he has inspired music. In 1990 singer-songwriter David Orr formed a rock band, Tusko Fatale, that acquired a minor cult following in the Virginia area, where it was based. Their song, "The Unfortunate Elephant," eulogizes Tusko: "The spider spins the more perfect web / The elephant, he drops over dead / The writer writes the more perfect line / In common we all lose track of time."

The scientific literature records no other cases of elephants given LSD. This makes Tusko and the two elephants in Siegel's 1982 experiment the sole pachyderm pioneers of the psychedelic experience.

Yet, paradoxically, elephants are widely associated with LSD. *Pink elephants* is a slang term for the drug, apparently inspired by a pattern commonly used on blotter paper during the 1990s. Then there is the scene from the 1941 Disney movie *Dumbo*, in which the big-eared elephant hallucinates enormous pink elephants on parade. Dumbo got high from drinking moonshine, not taking LSD, but users of the drug praise the scene as the perfect viewing material for a psychedelic trip. At the very least, it is a far more humane way to experience elephants on acid than the alternative.

———

West, L. J., C. M. Pierce, & W. D. Thomas (1962). "Lysergic Acid Diethylamide: Its Effects on a Male Asiatic Elephant." New Series, *Science* 138 (3545): 1100–3.

Racing Roaches

The starting light switches on. The gate opens, and the sprinter is off. She scurries directly onto the runway. The crowd in the stands goes wild. They wave their antennae back and forth, climbing all over one another in their excitement. Of course, most of them don't seem to be paying any attention to the sprinter, so perhaps it's the powerful floodlight that's causing them to go wild. Nevertheless, energized by their presence, the sprinter hurries forward, speeding toward the darkness looming ahead.

During the late 1960s psychologist Robert Zajonc spent a lot of time racing cockroaches—female *Blatta orientalis*, to be specific—in his University of Michigan lab. He carefully timed their runs with a stopwatch, and even built them a miniature stadium.

The stadium consisted of a 20 × 20 × 20-inch clear plastic cube. A runway, made of a transparent tube, ran straight through the cube, from a starting box at one end to a darkened goal box at the other. Zajonc also placed clear plastic boxes (aka bleachers) on either side of the runway. Cockroaches crawled around in these boxes—a captive audience.

Zajonc kept his roach-runners alone in a dark jar for a week before the big day, feeding them sliced apples, priming them for peak performance. On the day of the event, he placed a single roach in the starting box of the apparatus, turned on a 150-watt floodlight behind the box, and opened a gate, allowing the roach access to the tube. Away the roach would go, fleeing from the bright light toward the comforting darkness of the box at the other end.

Zajonc wasn't doing this for sport. At least, if he was, he didn't admit it. He was attempting to determine whether the athletic performance of a cockroach would improve in front of an audience of its peers—which is why he tested how fast the roaches ran both with an audience and without one.

What he found was that roaches definitely ran faster in the presence of other roaches. This fact alone might have merited his study a brief mention in a reference guide to cockroach behavior, but Zajonc argued his discovery had wider significance.

The phenomenon at work, he suggested, was "social facilitation." The mere presence of other cockroaches somehow gave his runners an extra boost of energy. And if this is true for cockroaches, then it might, he theorized, be true for humans. As Zajonc put it, "The presence of others is a source of nonspecific arousal. It can energize all responses likely to be emitted in the given situation." Translation: You'll probably run faster in front of a crowd than without one.

Why would this be? Zajonc chalked it up to an automatic,

physiological reflex. A creature that is alone can relax, but if members of its species are around, it needs to be more alert, in case it needs to respond to something they do. This extra alertness can enhance performance on a task such as running in a straight line. But there's a catch. The enhancement effect only works for simple tasks. The performance of complex tasks—ones that require some concentration—suffers in the presence of others. The extra energy creates sensory overload, making it harder to sort through thoughts.

Zajonc demonstrated this by adding a twist to his experiment. He made the roaches navigate a simple maze before they could reach the safety of the dark box. Sure enough, the roaches performed slower when challenged with figuring out the maze as their companions watched. To imagine how this observation might apply to humans, think of a task that requires some thought—solving mathematical problems, perhaps. Zajonc would predict that the effect of feeling self-conscious would inhibit performance of such a task in front of an audience.

Since Zajonc's experiment, the phenomenon has been tested in numerous species, including chickens, gerbils, centipedes, goldfish, and, of course, humans. The mere presence of others does, almost invariably, increase the speed of simple tasks and decrease the speed of complex tasks. In one test to confirm this, researchers used a telephoto lens to spy on joggers, attempting to discover whether they ran faster as they passed observers sitting by the side of the road. They did. The social-facilitation effect even appears to extend to mannequins. Subjects in a study at the University of Wisconsin performed a simple task faster when sitting in a room with a mannequin than when sitting in the room alone. Just about the only thing that hasn't been studied is whether the mere presence of a cockroach has a facilitating effect upon human performance. Judging by how fast some people run when they see the little creatures pop up in the kitchen, antennae waving, it probably does.

Zajonc, R. B., A. Heingartner, & E. M. Herman (1969). "Social Enhancement and Impairment of Performance in the Cockroach." *Journal of Personality and Social Psychology* 13 (2): 83–92.

Eyeing an Ungulate

Their eyes lock in a stare. It is a battle of wills. Who will look away first? The man focuses all his mental energy on his opponent. His eyes are like twin laser beams, intense and unwavering. His opponent stares back, her eyes black and impassive. For five long seconds they hold each other's gaze, until abruptly the sheep looks away. She bleats once and urinates on the ground.

Have you ever caught a stranger staring at you? Did it make you feel nervous, uncomfortable, or afraid? Now imagine you were a sheep. How do you think you would feel if a human was staring at you and wouldn't stop? That was what researchers at New Zealand's Massey University wondered.

To find out the answer, one of them stood inside a wooden-floored arena and stared at a sheep for ten minutes. He followed the animal's every movement with his eyes. The study's design protocol specified that "if the test sheep made eye contact, it was maintained until the sheep looked away." Thankfully for the pride of all involved, no instances were recorded in which the sheep stared down the human.

Throughout November 2001, a total of twenty sheep were subjected to the staring test. Next, the researchers repeated the experiment, substituting, in place of the staring man, either a cardboard box or a guy who gazed at the floor. Hidden observers recorded the behavior of the sheep in each situation. They looked, in particular, for signs of fear, such as freezing in place, attempting to escape, or glancing repeatedly at the stimulus.

The final data offered both expected and unexpected results. As expected, "individual sheep showed more fear- or aversion-related behavior in the presence of a human than with a cardboard box."

But, unexpectedly, the sheep displayed less fear toward the staring man than they did toward the eyes-downcast man. The sheep kept their distance from the guy staring at the floor, as if to say, "You won't look at me, so I don't trust you." However, they did urinate more often when directly stared at. The reason for that was unknown.

The Massey University study might seem a little eccentric, but there's actually a wealth of research involving the reactions of animals to staring humans. Animals that researchers have spent time staring at include iguanas, snakes, gulls, sparrows, and chickens. Such studies are part of a larger effort to understand predator-prey relationships—specifically, how prey animals react to visual cues from predators.

Researchers have also tested the reactions of humans to being stared at, although these studies get published in social-psychology journals rather than animal-behavior ones. In a widely cited 1972 experiment, Stanford University researchers pulled up to a red light on a motor scooter and fixedly stared at the driver of the car next to them. About four feet typically separated the experimenter and subject. Most of the people who were stared at exhibited a similar response. They would notice the staring scooter operater almost right away and then:

> Within a second or two, they would avert their own gaze and begin to indulge in a variety of apparently nervous behaviors, such as fumbling with their clothing or radio, revving up the engines of their cars, glancing frequently at the traffic light, or initiating animated conversation with their passengers. If there was a long time interval before the light changed, the subjects tended to glance furtively back at the experimenter, averting their gaze as soon as their eyes met his.

The experimenters stared until the light turned green, and then timed how quickly the drivers crossed the intersection. The drivers typically jammed down on the gas and flew through the lights. The researchers concluded that we humans interpret staring as a threat display and respond with avoidance behavior, such as speeding away from staring weirdos on motor scooters, to remove ourselves from danger.

Comparing the sheep and human studies, you might conclude that sheep are braver than humans. After all, the sheep didn't try to flee the presence of the staring human. However, it probably has nothing to do with bravery. A more likely interpretation is that sheep are so used to humans acting strange that they don't really care what we do anymore. And we humans can take comfort in the knowledge that at least we don't urinate when stared at. Or do we? The Stanford researchers had no way to observe whether that particular response occurred in any of their subjects. So if you ever decide to stare at another driver at an intersection and they suddenly get a sheepish grin on their face, you now might wonder why.

Beausoleil, N. J., K. J. Stafford, & D. J. Mellor (2006). "Does Direct Human Eye Contact Function as a Warning Cue for Domestic Sheep (Ovis aries)?" *Journal of Comparative Psychology* 120 (3): 269–79.

Lassie, Get Help!

Timmy has fallen down a well. "Lassie, get help!" he calls up from the darkness. Lassie pricks up her ears, looks down the well, and then takes off running. Soon she finds a ranger.

"Bark! Bark! Bark!"

"What is it, Lassie?" he says. "What are you trying to tell me?"

"Bark! Bark!" Lassie motions with her snout, then begins running back toward the well. Concerned, the ranger follows closely behind.

If you were trapped down a well like Timmy, what would your dog do? Would it run to get help, or would it wander off to sniff a tree? If you own a trained rescue dog it would probably get help, but what about an average dog, the kind whose greatest passions in life are (a) bacon, and (b) barking at the neighbor's cat? Would it figure out what to do in an emergency situation?

To find out, researchers Krista Macpherson and William Roberts from the University of Western Ontario arranged for twelve dog owners to pretend to have a heart attack while walking their dogs through an open field. The owners all performed the exact same actions. When they reached a pre-designated point in the field, marked by a target painted on the ground, they began breathing heavily, coughed, gasped, clutched their arm, fell over, and then lay motionless on the ground. A video camera hidden in a tree recorded what their dogs did next. In particular, the researchers were curious to see whether the dogs would seek help from a stranger sitting ten meters away.

The dogs—from a variety of breeds, including collies, German shepherds, rottweilers, and poodles—didn't do much to promote the theory of canine intelligence. They spent some time nuzzling and pawing their owners before taking the opportunity to roam around aimlessly. Only one dog—a toy poodle—directly made contact with the stranger. It ran over and jumped in the person's lap—not because it was trying to signal that its owner was in distress, but because it wanted to be petted. It probably figured, *Uh-oh! My owner's dead. I need someone to adopt me!*

Concerned that the heart-attack scenario may have been too subtle for the dogs—perhaps they thought their owners were just taking a nap—and that the presence of the passive stranger might have suggested to the dogs that nothing was wrong, the researchers designed a second, more dramatic test.

They arranged for each of fifteen dog owners to bring their dogs into an obedience school, greet a person in the front lobby, and then walk into a second room, where a bookcase then fell on the person. (Or, at least, the bookcase appeared to fall on the person. In reality, the researchers had shown each dog owner how to pull the piece of furniture down in such a way that it would only look like an accident without actually hurting the person.) Pinned beneath the shelves, each owner let go of his or her dog's leash and began imploring the animal to get help from the person in the lobby.

Once again, the canine response to the emergency was somewhat lacking. The dogs spent a good deal of time standing by their owners, wagging their tails, but not a single one went to get help. The researchers concluded that "the fact that no dog solicited help from a bystander—neither when its owner had a 'heart attack' nor when its owner was toppled by a bookcase and called for help—suggests that dogs did not recognize these situations as emergencies and/or did not understand the need to obtain help from a bystander." In other words, don't expect Fido to save your life.

The researchers were quick to point out that in some cases, dogs clearly have saved their owners' lives by seeking help. The media loves to report these stories, since they provide feel-good tales to end news broadcasts with—"Stay tuned for the dog that dialed 911!" But the researchers argue that such stories should not be considered indicative of typical dog behavior. So much for the urban legend of the life-saving pooch.

And while we're on the subject of urban legends, here's another one. "Timmy fell down a well" is perhaps the most quoted line from the *Lassie* TV show. So much so that Jon Provost, the actor who played Timmy, titled his autobiography *Timmy's in the Well*. However, although Timmy endured many calamities during the show—including falling into a lake, getting caught in quicksand, and being struck by a hit-and-run driver—he never once fell down a well.

Macpherson, K., & W. A. Roberts (2006). "Do Dogs (Canis familiaris) Seek Help in an Emergency?" *Journal of Comparative Psychology* 120 (2): 113–19.

Horny Turkeys and Hypersexual Cats

Male turkeys aren't fussy. They'll try to mate with almost anything, including a head on a stick.

During the late 1950s Martin Schein and Edgar Hale, animal researchers at Pennsylvania State University, observed that when they placed male turkeys in a room with a lifelike model of a female turkey, the birds responded sexually to the model in a manner "indistinguishable from their reaction to receptive females." The turkeys let out an amorous gobble announcing their intentions, began waltzing around and puffing up their feathers, and finally mounted the model and initiated a copulatory sequence.

Intrigued, Schein and Hale wondered what the minimal stimulus was that would elicit a sexual response from a male turkey. Specifically, how many of the model's body parts could they remove before the turkey would lose interest? As it turned out, they were able to remove quite a few.

Tail, feet, and wings—they all proved unnecessary. Finally the researchers gave the turkeys a choice between a headless body and a head-on-a-stick. To their surprise, the turkeys invariably chose the head-on-a-stick over the body. Apparently this was all it took to get a turkey going. The researchers wrote:

A bodyless head supported 12 to 15 inches above floor level elicited courtship display, approach, and mounting movements properly oriented behind the head . . . The male's subsequent response to the model head included treading movements, tail lowering, and movements directed towards achieving cloacal contact. At this point ejaculation could be evoked by applying mild tactile stimulation to the exposed penile papillae of the male. Effective stimuli included a warm watchglass, a human hand, or any warm smooth surface.

The male's fixation on the female head appeared to stem from the mechanics of turkey mating. When a male turkey mounts a female, he is so much larger than her that he covers her completely, except for her raised head. Therefore, it is her head alone that serves as his erotic focus of attention.

Having isolated the head as the ultimate turkey turn-on, Schein and Hale next investigated how minimal they could make the head before it failed to elicit a response. They tested turkeys with a variety of heads-on-a-stick—a fresh female head; a fresh male head; a two-year-old female head that was "discolored, withered, and hard"; a similarly dried-out male head; and a series of balsa-wood heads that varied with respect to the presence of eyes and beaks.

The fresh female head was the clear winner, followed by the dried-out male head, the fresh male head, and the old female head. All the wooden heads came in a distant last place—indicating, perhaps, that turkeys prefer women with brains. But it should be noted that even though the wooden heads were not a preferred object of passion, they nevertheless did elicit sexual behavior.

Curious about the mating habits of other poultry, Schein and Hale performed similar tests on white leghorn cocks. They discovered that these birds, unlike turkeys, required a combination of head and body for adequate sexual stimulation. They detailed this research in a paper with the evocative title, "Effects of Morphological Variations of Chicken Models on Sexual Responses of Cocks."

Poultry are not the only birds easily misled in matters of romance. Konrad Lorenz once observed a shell parakeet who grew amorous with a small celluloid ball. And many other animals exhibit mating behavior toward what researchers refer to as "biologically inappropriate objects." Bulls will treat almost any restrained animal as a receptive cow. Their general rule in life seems to be, in the words of Schein and Hale, "If it doesn't move away and can be mounted, mount it!"

During the early 1950s researchers at Walter Reed Army Medical Center surgically damaged the amygdala (a region of the brain) in a number of male cats. These cats became hypersexual, attempting to mate with a dog, a female rhesus monkey, and an old hen. Four of these hypersexual cats, placed together, promptly mounted one another.

At the summit of this beastial pyramid of the perverse stands *Homo sapiens*. We, as a species, are in no position to laugh at the mating habits of turkeys, bulls, or any other creature, since there are no apparent limits to what can serve as an erotic stimulus for a human being.

Case in point—Thomas Granger, a teenage boy who in 1642 became one of the first people to be executed in Puritan New England. His crime was having sex with a turkey (as well as a few other animals). Granger confessed to the deed, but defended himself by arguing that sex with animals was a custom "long used in old England." Ah yes, Merrie Olde England! For some reason, this story—though entirely true—seldom makes its way into history books. Nevertheless, it's an interesting piece of trivia to weave into the conversation during Christmas dinner.

Schein, M. W., & E. B. Hale (1965). "Stimuli eliciting sexual behavior." In *Sex and Behavior* (F. A. Beach, ed.). New York: John Wiley & Sons.

The Brain Surgeon and the Bull

The matador stands in the bullring, the hot Spanish sun beating down on his head. Thirty feet away stands a bull. A hushed crowd watches in the stands. The matador flourishes his red cape. The bull stamps one hoof, snorts, and then charges. At the last moment, the matador lifts the cloth and gracefully steps to the side, but the bull unexpectedly swings its head and knocks the matador to the ground. The crowd gasps. People start screaming as the bull circles around and charges again, bearing down on the matador, who lies clutching his stomach. It seems like a bad situation for the matador. The bull is only yards away. In a few seconds it will all be over. But suddenly the matador reaches to his side and presses a button on his belt. Instantly a chip inside the bull's brain emits a small burst of electricity, and the bull skids to a halt. It huffs and puffs a few times, then passively walks away.

Could this be how bullfights of the future will be fought, with brain-control devices implanted in bulls to prevent injuries to the matadors? It might lead to a drop in ticket sales, but it could also save lives. However, there's no need to wait for the future. Such technology has existed for decades. The ability to use brain chips to stop bulls was first demonstrated by Yale researcher José Delgado in 1963.

Delgado was a Spanish-born researcher who accepted a position at Yale University's School of Medicine in 1950. Over the next two decades he became one of the most visible members of a generation of researchers who pioneered the science of Electrical Stimulation of the Brain (ESB).

ESB involves using wires implanted inside the skull to

stimulate different regions of the brain. Such stimulation can produce a wide variety of effects, including the involuntary movement of limbs, the eliciting of emotions such as love or rage, or the inhibition of appetite and aggression. To its critics, it has always smacked of Orwellian mind control, but its defenders insist this is a misconception, pointing out that it's difficult to predict exactly what will happen when a specific region of the brain is stimulated. Nor is it possible to control thoughts or complex forms of behavior.

Delgado's great innovation was to invent an ESB chip with a remote-control unit. He called the chip a stimoceiver. It allowed him to study subjects as they moved around in a natural way, free of cables dangling from their heads. Even though TV remote controls were not yet in widespread use during the 1960s (and were the size of a brick), this was the metaphor he used to describe his invention:

> The doors of a garage can be opened or closed by pushing a button in our car; the channels and volume of a television set can be adjusted by pressing the corresponding knobs of a small telecommand instrument without moving from a comfortable armchair . . . These accomplishments should familiarize us with the idea that we may also control the biological functions of living organisms from a distance. Cats, monkeys, or human beings can be induced to flex a limb, to reject food, or to feel emotional excitement under the influence of electrical impulses reaching the depths of their brains through radio waves purposefully sent by an investigator.

With the stimoceiver, most of Delgado's experiments followed a similar pattern. He would stimulate different regions of a subject's brain and then observe what happened. He described much of this research in his 1969 book *Physical Control of the Mind: Toward a Psychocivilized Society*. One of his more sensational experiments involved a monkey called Ludy. By pressing a button, he caused her to perform a complex sequence of actions that included turning her head to the right, standing up, circling to the right, climbing a pole, descending to the floor, uttering a growl, threatening a subordinate monkey, and then returning to the monkey group in a peaceful manner. She performed this behavior twenty thousand times in a row.

Delgado could use his remote-control unit to manipulate human subjects just as easily as animal ones. He made one patient repeatedly clench his fist against his will, until the man said, "I guess, Doctor, that your electricity is stronger than my will." Anxiety, rage, and love could also be dialed up on command. He rigged up a knob that he used to increase or decrease the amount of anxiety a female patient experienced. He pressed a button and caused another patient to fly into such a rage that she hurled a guitar across the room. Yet another patient violently shredded sheets of paper even as she moaned, "I don't like to feel like this." Other patients were luckier and got to experience the love button. When stimulated in this way, two patients became so overwhelmed that they expressed a desire to marry the researcher. One was a thirty-six-year-old woman. The other was an eleven-year-old boy.

But the experiment that Delgado will forever be remembered for is his time in the bullring. It took place in 1963 at a bull ranch in Cordova, Spain. He implanted electrodes into the brains of "brave bulls"—bulls that were known for their aggressive tendencies. He then tested the effects of ESB on these creatures. He found he could make the bulls turn their heads, lift one leg, or walk around in a circle. In addition, "vocalizations were often elicited, and in one experiment to test the reliability of results, a point was stimulated 100 times and 100 consecutive 'moo's' [sic] were evoked." It was kind of like a real-life version of Big Mouth Billy Bass—or Billy Bull, one should say.

As a final test, Delgado got into a ring with one of the bulls. He stood by the side of the arena and waved a red cloth at the beast. The bull began to charge. When it was mere feet away from him, Delgado pressed a button and the bull abruptly stopped. Delgado admitted to worrying that the stimoceiver might choose that moment to malfunction, but everything worked perfectly. Delgado suggested the stimulation caused a sudden inhibition of the bull's aggression, but other researchers argue it probably simply caused the bull to turn sharply to the side, frustrating the animal's ability to charge. Delgado has acknowledged this may have been the case. Whatever the stimoceiver did, it worked.

The media gave Delgado's bull experiment widespread coverage—including a front-page story in the *New York Times* —making him an instant scientific celebrity. However, rival brain researchers were less impressed. Elliot Valenstein, a professor at the University of Michigan, criticized Delgado, claiming, "His propensity for dramatic, albeit ambiguous, demonstrations has been a constant source of material for those whose purposes are served by exaggerating the omnipotence of brain stimulation."

During the 1970s and '80s ESB came under attack from those who feared it would be used to create a totalitarian state of mind-controlled zombies. Funding dried up, and Delgado moved back to Spain, where he focused on noninvasive methods of brain research. However, during the past decade, interest in ESB has revived, thanks to advances in computers and electronics, as well as to the recognition that brain chips can offer enormous therapeutic benefit to patients suffering from disorders such as epilepsy, Parkinson's disease, depression, or chronic pain.

Scientists have also been reviving the research tradition, started by Delgado, of designing remote-controlled animals. John Chapin, a professor at the State University of New York in Brooklyn, has designed a camera-equipped, remote-controlled rat that he hopes will be able to help rescue workers find survivors in rubble piles after earthquakes and other disasters. A Chinese researcher made headlines in 2007 with a remote-controlled pigeon, and there have been reports of the Pentagon funding research into the creation of a remote-controlled shark that could be used to track boats without detection. Fisheries have even been investigating the possibility of inserting neural implants into farmed fish. The fish could be allowed to roam free in the ocean and then, with a push of a button, be called back to be harvested. But, of course, it is the scientist who can create a remote-controlled dog with a "fetch beer" button, who will doubtless become the first ESB billionaire.

Delgado, J. M. R. (1969). *Physical Control of the Mind: Toward a Psychocivilized Society*. New York: Harper & Row.

CHAPTER SIX

Mating Behavior

The public has long maintained a gossipy interest in the sex lives of scientists. Muckraking biographers whisper that Isaac Newton may have died a virgin, that Nikola Tesla was devoutly celibate, that Albert Einstein had a mistress, and that Richard Feynman was a ladies' man. However, for much of history scientists steadfastly refused to develop a counter-vailing interest in the sex lives of the general public. Writing in 1929, the psychologist John Watson bemoaned the lack of scientific knowledge about sex: "The study of sex is still fraught with danger . . . It is admittedly the most important subject in life. It is admittedly the thing that causes the most shipwrecks in the happiness of men and women. And yet our scientific information is so meager. Even the few facts that we have must be looked upon as more or less bootlegged stuff."

In the decades following Watson's complaint, scientists did begin to undertake serious studies of human sexuality. The most famous researchers in this field were Alfred Kinsey, William Masters, and Virginia Johnson. However, even as attitudes toward such work grew less restrictive, the research remained more descriptive than experimental. As late as 1989, the psychologists Russell Clark and Elaine Hatfield observed that "until recently, scientists have had to rely almost exclusively on interviews and naturalistic studies for their information [about human sexuality]. Only recently have researchers begun to conduct laboratory experiments."

Happily for us, while experimental studies of the mating behavior of humans may have been taboo throughout much of the history of science, there definitely were some—and in recent years the number has increased to quite a few. As you might expect, such experiments can be relied on to provide riveting reading.

1. ATTRACTION

Arousal on a Creaky Bridge

People who cross the Capilano Suspension Bridge do so very carefully. The narrow, 450-foot-long bridge, located just outside of Vancouver, British Columbia, is built of wood and wire cables. It has the unnerving tendency to sway, wobble, and creak in the wind, giving the impression it might flip over at any moment. For those standing on it, this would result in a 230-foot plummet onto the rocks below.

In 1974 an attractive woman approached single young men as they were crossing the bridge. Coyly gripping the low handrail to keep her balance, she asked if they would be willing to participate in a psychology experiment. She told them she was investigating "the effects of exposure to scenic attractions on creative expression." Once they agreed, she showed them a picture of a woman holding a hand over her face and asked them to write a brief, dramatic story about it. They did this while standing on the bridge, rocking in the wind, trying not to think about the drop below.

When they were done, the interviewer smiled and thanked them. Then, as though on the spur of the moment, she suggested they call her to learn more about her research. She wrote down her name, Gloria, and phone number on a scrap of paper and handed it to them.

Thirteen of the twenty men she interviewed later called. It was clear they weren't interested in her research. They were interested in her. But the woman wasn't on the bridge fishing for dates. Nor was she there to explore the relationship between creative expression and scenic attractions. That was just her cover story. The true purpose of the experiment was to explore the link between fear and sexual arousal.

The experiment had been designed by two researchers, Donald Dutton and Arthur Aron. They hypothesized that men would find the interviewer more attractive, and thus be more likely to call her, if they met her in a fear-inducing environment such as the Capilano Bridge.

To confirm their theory, the researchers next had the woman approach men in a calmer environment—as they were relaxing on a park bench. She offered the same explanation, "This is a psychology experiment . . . " and again gave out her name and number. This time she identified herself as Donna.

Under these conditions, only seven of twenty-three men later called. Not a bad response rate—these were still young guys, after all—but not a terrific one. The content of the stories written by each group also showed a marked difference. The men on the bridge wrote stories containing far more erotic imagery. They seemed to be highly aroused by the encounter. These results suggested the existence of a link between sexual arousal and fear.

The experimenters had guessed fear would enhance sexual arousal because of the concept of "misattribution of arousal." When we're in a situation that prompts strong emotions, this theory suggests, we often mistakenly attribute those feelings to the person we're with, rather than to the situation. So if you're on a suspension bridge two hundred feet off the ground, your senses will be on high alert. Your heart will beat faster. Your palms may grow sweaty. Your stomach will wind into a knot. If an attractive woman (or man) then approaches you, you associate your feelings with the person. You think you're falling in love, when in reality you're just afraid of falling.

Would-be Romeos can take advantage of this insight. Scare your date! Take her to a horror movie, for a ride on a motorcycle, or on a stroll across a creaky suspension bridge. Just remember that ideally you want her to be frightened by the situation, not by you.

Dutton, D. G., & A. P. Aron (1974). "Some Evidence for Heightened Sexual Attraction under Conditions of High Anxiety." *Journal of Personality and Social Psychology* 30 (4): 510–17.

The Hard-to-Get Woman

Are women who play hard to get more desirable? Popular wisdom says so, as do many romance columnists. No less an authority than the Roman poet Ovid wrote, "Easy things nobody wants, but what is forbidden is tempting." Elaine and William Walster, of the University of Wisconsin, decided to put the idea to the test with the aid of a woman willing to play alternately hard or easy to get. Since she interacted with a lot of men in her profession, she seemed like the perfect guinea pig. She was a prostitute.

The experiment occurred at her place of business—a brothel in Nevada. The procedure was as follows: When each client arrived at her room, she mixed him a drink. Then she delivered the "experimental manipulation." Half the time this involved playing hard to get. She would tell her client, "Just because I see you this time it doesn't mean that you can have my phone number or see me again. I'm going to start school soon, so I won't have much time, so I'll only be able to see the people that I like the best." Then she proceeded to get to work.

The other half of the time, in the easy-to-get condition, the prostitute skipped the lecture and got right down to business.

The experimenters measured the clients' desire for the prostitute in a variety of ways. They asked her to rate how much each guy seemed to like her, and they recorded the amount of money paid by the clients and the number of times during the following month the men returned.

When all the data was collected, the results were clear. The hypothesis that a hard-to-get prostitute would be more desirable was flat-out wrong. Clients were far less likely to return after being warned off. Apparently men who visit prostitutes don't like them to be too fussy.

In subsequent experiments conducted at a dating service, the researchers corroborated that these results applied to all romantic interactions, not just to prostitute-client relationships. Men, despite what romance columnists say, do not like women who are uniformly hard to get. What they do like, the Walsters figured out, are women who are selectively hard to get—who are cold and standoffish to all the other guys, but warm and receptive to them. So what the prostitute could have said to ensure repeat business was, "I won't have much time because I'm going to start school soon, so I'll only be able to see the people I like the best. Which means I'll always have time for you." However, this strategy has not yet been experimentally confirmed.

Walster, E., G. W. Walster, J. Piliavin, & L. Schmidt (1973). "'Playing Hard To Get': Understanding an Elusive Phenomenon." *Journal of Personality and Social Psychology* 26 (1): 113–21.

Love at Last Call

Country music has doubtless made numerous contributions to the advancement of science, but it seldom has played a role as prominently as it did in James Pennebaker's study of mating psychology in bars. Pennebaker was listening to Mickey Gilley's classic song "Don't the Girls All Get Prettier at Closing Time," and he wondered, "Is that true? Do the girls really get prettier?" Determined to find out, he assigned teams of his students to visit bars near the Charlottesville campus of the University of Virginia one Thursday evening in 1977.

The two-member teams entered the bars at three separate times—nine o'clock, ten thirty, and midnight. They approached both male and female bar patrons who appeared to be alone, explained they were conducting a psychology study, and asked the patrons to rate the attractiveness of the other occupants of the bar on a scale of one to ten.

Gilley, it turned out, was right. Men did rate the women as more attractive as closing time neared. Likewise for the women's ratings of the men. But attractiveness ratings for members of the same sex decreased. (Evidently the teams visited heterosexual watering holes.)

Pennebaker suggested these results demonstrated "reactance theory"—as our time to make a decision lessens, we react by panicking and thinking all the choices look pretty good. So as bar patrons run out of time to decide whom to go home with, all possible romantic partners begin to seem equally compelling. However, Pennebaker conceded the results could also be an effect of increased alcohol intake skewing people's judgment as the night progressed. This is popularly known as the beer-goggle phenomenon.

Pennebaker's methodology and reasoning seemed sound, but subsequent attempts to replicate his study produced mixed results. A 1983 study obtained similar results at a blue-collar country-and-western bar in Georgia, but not at a college bar. A 1984 study in Madison, Wisconsin, entirely failed to replicate Pennebaker's results. It found that ratings of perceived attractiveness *declined* as closing time approached, especially the ratings women gave to men. Nor did alcohol intake appear to have an effect on ratings.

In 1996 a study in Toledo, Ohio, attempted to resolve these disparate findings by asking more specific questions of bar patrons. It found the closing-time effect did exist, but only for people who specified they were single. People in a relationship apparently did not feel the same pressure to lower their standards as the minute hand neared last orders, because they knew they didn't need to make a choice. They were already taken, so to speak. The researchers warned, "One important implication of these findings is that those who should be discriminating in choosing a partner (i.e., those not in a relationship) may make unwise and later regrettable choices." This doesn't come as news to all the battle-scarred veterans of the late-night dating scene.

Pennebaker, J. W., M. A. Dyer, R. S. Caulkins, D. L. Litowitz, P. L. Ackreman, D. B. Anderson, & K. M. McGraw (1979). "Don't the Girls Get Prettier at Closing Time: A Country and Western Application to Psychology." *Personality and Social Psychology Bulletin* 5 (1): 122–25.

The Gay Detector

The experiment began innocently enough. Pairs of Stanford undergraduate men, all of whom had volunteered to participate in a psychological study, were taken to a room. After meeting each other, the two people sat together at a table facing a projection screen. In front of each of them was a dial. The researcher, Dana Bramel, explained that the dial displayed the output of a "psychogalvanic skin response apparatus."

Now things got a little stranger. The apparatus, they were told, would be measuring their subconscious responses to a series of pictures. The pictures would show men in partial stages of undress. Bramel emphasized the next point—any movement of the dial indicated homosexual arousal on the part of the subject. Strong homosexual feelings would cause the dial to "go off the scale." But to put the undergraduates at ease, he assured them they alone could see the dial in front of them, and all data would be kept anonymous.

Finally, Bramel informed them of their task. They were to record whatever number their dial reached, and then predict the score of their fellow participant. The challenge was essentially to guess how gay the other person was.

The slideshow began. Bramel hadn't been kidding when he said the photos would show partially undressed men. The models were, in fact, almost nude, and posed in various seductive stances. Each subject looked at the photos and then looked down at the dial in front of him. The dial was twitching vigorously, as though it had acquired a life of its own. The subject began to realize that, according to the psychogalvanic skin response apparatus, he was getting turned on. As the slideshow progressed and the pictures grew more risqué, the needle pulsed upward ever more energetically, like a finger wagging accusingly, saying, *Naughty boy, you're having dirty thoughts!* The machine was telling him he was gay.

In the modern metrosexual age, a Stanford undergrad might easily laugh off learning he has latent "homosexual tendencies." He might even readily admit to such inclinations. But this experiment did not occur post–*Will & Grace.* It took place in the late 1950s, at a time when homosexuals were one of the most stigmatized and openly discriminated-against groups in American society. Homosexuality was a legal reason for firing someone, for refusing him service, or for throwing him in prison. The American Psychiatric Association even listed homosexuality as a form of mental illness. Therefore, being flagged as gay was not something an undergrad could easily shrug off.

So as the subject looked at the surging needle on his dial, it must have been with a mixture of dismay and terror. He quietly recorded his scores and tried to guess what the guy next to him had rated. His goal was probably to finish the experiment as soon as possible and get out of there. But he actually had no reason to fear. He wasn't about to be outed as gay. The readings on the dials were all the same, and all completely bogus. The needles were secretly controlled by the experimenter, who was cranking them up to higher levels as the pictures grew more graphic. Only eight participants ever suspected this. The others, all ninety of them, fell for the ruse completely. Bramel noted, "The expression of relief which often followed the unveiling of the deceptions indicated that the manipulations had been effective."

Bramel's true purpose was to determine how people would react to learning they possessed a socially stigmatized quality such as homosexuality. Would they project the same 'negative' quality onto other people, in a defensive attempt to get others into the same boat? The answer was yes. That's exactly what they did. When predicting the score of the other guy in the room, subjects tended to guess it was similar to their own. This was especially true for people who had been judged before the experiment to have high self-esteem.

However, this finding is not the primary reason Bramel's experiment is remembered today. Instead, the study has become a frequently cited example in the debate over when deception in psychological research goes too far. Critics of the experiment insist that convincing unsuspecting undergrads they were closet homosexuals went well beyond any scientific justification. Imagine, in particular, the fear of any participant who actually was gay, as he sat watching the twitching dial threaten to expose his secret to the world.

Of course, where science now hesitates to go, reality TV has no qualms about rushing in. In 2004 Fox TV produced *Seriously, Dude, I'm Gay*, in which heterosexual contestants competed to fool their friends and family into believing they were gay. Because of widespread public complaints, the show got canned before making it to air. But it can't be long before studio executives, in their wisdom, decide to repackage and air Bramel's experiment as a reality-based game show. When they do, an appropriate title might be *Seriously, Dude, You're Gay.*

———

Bramel, D. (1962). "A Dissonance Theory Approach to Defensive Projection." *Journal of Abnormal and Social Psychology* 64 (2): 121–29.

2. SEX

Heart Rate During Intercourse

In 1927 doctors Ernst Boas and Ernst Goldschmidt invented the cardiotachometer. This medical instrument allowed physicians to do something never before possible—measure the heart rate nonintrusively for prolonged periods of time. By modern standards the cardiotachometer was actually quite intrusive. Two rubber straps held copper electrodes to the chest, and these electrodes, in turn, were attached to a one-hundred-foot wire that led to a room full of recording equipment. Being attached to the cardiotachometer was like being on a long leash. But as long as people ignored the cord, they could do everything just as they normally would. So, like kids with a new toy, Boas and Goldschmidt set out to measure the heart rate during every activity imaginable.

They recruited volunteers to come to New York's Mount Sinai Hospital and have their heart rates monitored going about the actions of daily life—standing, walking, exercising, dancing, sitting, talking, and eating. They observed people playing poker and discovered the cardiotachometer was a useful bluff detector. "One poker player, in particular," they recorded, "invariably showed a brief acceleration of the heart when he held a good hand."

They monitored people going to the bathroom: *Enters bathroom—86, defecating—89, washing hands—98.*

And, in an effort to leave no stone unturned, they measured the heart rate of a husband and wife during intercourse. The results were illuminating. During orgasm the heart rate rose to 148.5. This exceeded the rate of any other recorded activity, including strenuous exercise. "The curve of heart rate clearly indicates the strain placed on the cardiovascular system," they warned, "and helps to explain some cases of sudden death during and after coitus."

But they noted something else: "The record illustrated in Figure 47 was taken on the first evening. It shows four peaks of heart rate for the woman, each peak representing an orgasm." Four orgasms! The researchers make no more mention of this, as if to say *Nothing to see here, keep moving along.* But clearly the rubber straps, tether, and audience in the next room hadn't detracted from the wife's experience.

Almost thirty years would pass before another researcher, Dr. Roscoe G. Bartlett, next recorded the heart rate during intercourse. When he did, he took far more elaborate steps than his predecessors to protect the modesty of everyone involved. He refused to disclose the location of the experiment, or the identity of the subjects (three couples). He himself did not know their identities—only an intermediary did. Between the time each couple arrived at the research facility, entered a private room, attached electrodes to themselves, pressed the appropriate buttons to indicate the occurrence of "intromission, orgasm and withdrawal," and subsequently left the building, they were never seen by the investigators. Bartlett's three female subjects did not exhibit the same orgasmic frequency as the 1928 subject. Perhaps these efforts at modesty were to blame.

Bartlett subsequently left science and embarked on a profession that has placed him in more intimate contact with sex and sleaze. Since 1993 he has been serving in the U.S. Congress as a Republican representative from Maryland.

Boas, E. P., & E. F. Goldschmidt (1932). *The Heart Rate*. Baltimore, MD: Charles C. Thomas.

The Multiorgasmic Male

Dr. Beverly Whipple specialized in studying the health and sexuality of women. However, when, in the late 1990s, a man contacted her claiming to have a unique physical ability, her curiosity was piqued. The man—never named by Whipple— stated he was able to have multiple orgasms.

Many women are able to have multiple orgasms. Doctors Boas and Goldschmidt, as we just read, encountered such a woman in their 1928 study. But after a man experiences an orgasm his body is flooded by hormones that shut down the sexual response, making it temporarily impossible for him to achieve orgasm again. Earlier researchers had recorded a handful of cases in which men taught themselves the ability to suppress ejaculation, thereby allowing themselves to have multiple orgasms, but the talent of Whipple's correspondent went beyond this. He claimed he could have an orgasm with ejaculation, remain erect, and then promptly have another orgasm—time after time.

Whipple invited the mystery man to visit the human physiology laboratory in the College of Nursing at Rutgers University, where she was a professor. The experiment she had in mind was simple. She challenged him to sit in the lab and perform his trick for as long as possible.

The man did not disappoint. He proceeded to bring himself to orgasm six times with six ejaculations in a period of thirty-six minutes. He maintained an erection throughout the entire process, despite the discomfort of the laboratory conditions. There was a blood-pressure cuff on his arm, a heart-rate monitor on his big toe, and an infrared camera in front of his face, constantly scanning the diameter of his pupil. The research team observed him through a window. There was also no air-conditioning in the room, which was the reason he gave for stopping at six. He said it had grown too hot and stuffy. He swore that in a more comfortable environment he could have carried on to ten, or more.

Whipple reports that since she published this study, numerous men have contacted her claiming the same ability. While this is of scientific interest, the existence of such over-achievers offers little comfort to the vast majority of men who do not enjoy such a talent.

———
Whipple, B. (1998). "Male Multiple Ejaculatory Orgasms: A Case Study." *Journal of Sex Education and Therapy* 23 (2): 157–62.

Pushing the Pleasure Button

Patient B-19 was a troubled young man. After dropping out of high school, he drifted through a variety of dead-end jobs— stock clerk, janitor, factory worker. Eventually he ended up in the military—and was promptly kicked out for displaying homosexual tendencies. He then fell into a life of vagrancy, drug use, and homosexual prostitution. When he came to the attention of Dr. Robert G. Heath of Tulane University, he was complaining of feelings of apathy, boredom, inferiority, depression, and alienation. He confessed he frequently contemplated suicide.

Heath took one look at B-19 and saw the perfect candidate for an experiment he had been contemplating. Heath had been researching electrical stimulation of the brain. In particular he had focused on the septal area, which, when stimulated, triggers feelings of intense pleasure and sexual arousal. Heath wondered whether he could use this to change a person's behavior. Could he transform a homosexual man, such as B-19, into a heterosexual?

The peculiar pleasure-inducing properties of the septal region had been discovered by James Olds and Peter Milner of McGill University in 1954. They had placed an electrode inside a rat's brain and noticed that when they delivered a small shock to the septal region, the rat seemed to enjoy the sensation. In fact, it craved it. The rat learned to press a lever to deliver a shock to itself, and soon it was pressing the lever two thousand times an hour. Rats would press the lever rather than eat. Mother rats would abandon their young to get a few jolts. Olds and Milner declared the septal area to be the brain's pleasure center.

Heath lost little time applying Olds and Milner's discovery to human subjects. He discovered that, in addition to a pleasure center, the brain had an "aversive system," a punishment center. By stimulating these regions he could temporarily turn a person into a homicidal maniac, or the happiest person in the world. He once inserted a thin tube into a woman's brain and delivered acetylcholine, a chemical stimulant, directly to her pleasure center; he recorded that she experienced intense orgasms for over half an hour.

The experiment meant to transform B-19 into a heterosexual took place in 1970. Heath implanted stainless steel Teflon-insulated electrodes, 0.003 inch in diameter, into the septal region of B-19's brain. Three months later, after B-19 had fully healed from the surgery, the conversion program began.

Heath first showed B-19 a pornographic film. An EEG measured brain activity as B-19 sat in the lab viewing the scenes of "sexual intercourse and related activities between a male and female." B-19 displayed no significant response. He sat passively. His brain waves showed only "low amplitude activity."

With B-19's homosexuality (or lack of interest in heterosexual pornography) thus confirmed, Heath commenced the septal stimulation. B-19 received a few minutes of therapy each day—mild shocks conducted into his brain. B-19 liked the sensation. He remarked that it was similar to the effect he felt when he took amphetamines. After a week of this, his mood had remarkably improved. He smiled more, looked relaxed, and reported stirrings of sexual motivation.

Heath then rigged up a device that allowed B-19 to push a button and deliver a one-second shock to himself. Giving him this power was like letting a chocaholic loose in a candy shop. During one three-hour session, B-19 pressed the button 1,500 times, approximately once every seven seconds. Heath noted, "During these sessions, B-19 stimulated himself to a point that, both behaviorally and introspectively, he was experiencing an almost overwhelming euphoria and elation and had to be disconnected." At the end of each session B-19 whined and begged not to have the pleasure button taken away: *Please, let me press it just one more time!*

By this stage in the therapy, B-19 was feeling pretty good. His libido had been so jacked up he was expressing sexual interest in just about everything, including the female nurses. When Heath showed him the stag film again, B-19 "became sexually aroused, had an erection, and masturbated to orgasm." Truly, a changed man.

Over the next few days, B-19's state of arousal grew ever stronger, and Heath decided it was time to take the next step. He would give B-19 an opportunity to have sex with a woman—something B-19 had never done before. All his prior sexual experiences had been homosexual.

After receiving permission from the state attorney general, Heath arranged for a twenty-one-year-old female prostitute to visit the lab. He warned the young woman the situation was going to be a little weird. Undeterred, she agreed to do it for fifty dollars. Heath draped black curtains around the lab to allow the couple some privacy. Some candles and a Barry White record would have really set the mood, but this was, after all, a place of science, not a bordello. Heath then prepared B-19 by giving him free rein with the self-stimulator. When good and buzzed, B-19 was ushered in to see the prostitute.

The prostitute must have thought "weird" was an understatement. B-19 was not only charged up from the self-stimulation, he had wires coming out of his head to allow the researchers to monitor his brain waves during the encounter. The research team waited in the room next door for the action to begin.

B-19 got off to a slow start. He spent the first hour nervously delaying, talking about his "experiences with drugs, his homosexuality and his personal shortcomings and negative qualities." At first the prostitute allowed him to take his time, but as the second hour began, she became restive. Evidently she didn't want to spend all day there, so to speed things up she stripped off her clothes and lay down next to him. At this point, Heath's description of the experiment, published in the normally rather unexciting *Journal of Behavior Therapy and Experimental Psychiatry*, suddenly turns into something akin to a letter from "Penthouse Forum":

> In a patient and supportive manner, she encouraged him to spend some time in a manual exploration and examination of her body, directing him to areas which were particularly sensitive and assisting him in the initial manipulation of her genitalia and breasts. At times, the patient would ask questions and seek reinforcement regarding his performance and progress, to which she would respond directly and informatively. After about 20 min of such interaction she began to mount him, and though he was somewhat reticent he did achieve penetration. Active intercourse followed during which she had an orgasm that he was apparently able to sense. He became very excited at this and suggested that they turn over in order that he might assume the initiative. In this position he often paused to delay orgasm and to increase the duration of the pleasurable experience. Then, despite the milieu and the encumbrance of the electrode wires, he successfully ejaculated.

Mission accomplished! B-19 was now, as far as Heath

was concerned, a bona fide heterosexual. A few days later Heath released the virile young man out into the world. Heath checked on his progress a year later, and noted with satisfaction that B-19's new heterosexual inclinations had apparently persisted because B-19 reported having had an affair with a married woman. Unfortunately, he also confessed to engaging in homosexual behavior twice, "when he needed money and 'hustling' was a quick way to get it." Nevertheless, Heath declared the experiment a success. He predicted "future effective use of septal activation for reinforcing desired behavior and extinguishing undesired behavior."

We know nothing of B-19's later fate. So it's unclear whether B-19 truly transformed into a heterosexual, or whether the entire experience was simply a one-off. The latter seems more probable.

Heath continued his work with septal stimulation, though he did not attempt any more conversions of homosexuals. During the remainder of the decade, he worked on developing a battery-powered "brain pacemaker." This device could deliver low levels of stimulation to the brain for extended periods of time. It showed great potential, especially for calming extremely violent or depressed patients. However, the medical community was reluctant to embrace Heath's work. It smacked too much of mind control. Only in recent years has interest in electrical stimulation of the brain gained momentum again. (See chapter five, "The Brain Surgeon and the Bull," for examples.)

More surprisingly, the implantation of septal electrodes for purely recreational use never caught on—despite the predictions of some, such as Timothy Leary, that it would soon be all the rage. It's probably just as well this hasn't become the stimulant of choice. Would the world be ready for pleasure at the push of a button?

Moan, C. E., & R. G. Heath (1972). "Septal Stimulation for the Initiation of Heterosexual Behavior in a Homosexual Male." *Journal of Behavior Therapy and Experimental Psychiatry* 3: 23–30.

Voulez-Vous Coucher Avec Moi (Ce Soir)?

It's a sunny day. A young guy is walking alone on a college campus, minding his own business. He's reasonably attractive, though he has never considered himself a babe magnet. But suddenly a good-looking woman stops him and says, "I have been noticing you around campus. I find you to be attractive. Would you go to bed with me tonight?" He looks at her, startled but intrigued. "Well, isn't this my lucky day," he thinks.

Actually, it wasn't his lucky day. The poor guy was just an unwitting subject in an experiment conducted in 1978 on the campus of Florida State University.

It all started in the classroom of Russell Clark, during a meeting of his course on experimental social psychology. Clark was discussing James Pennebaker's Don't the girls get prettier at closing time? experiment—the same one described just a few pages earlier. He made an offhand comment to the effect that only guys need to worry about honing their pickup lines; women can just snap their fingers and men come running. Some of his female students took exception to this generalization, so Clark issued a challenge: Let's put it to an experimental test. Let's find out in a real-life situation which gender would be more receptive to a sexual offer from a stranger. The students took him up on the challenge, and a bizarre experiment was born.

Nine of Clark's students—five women and four men—fanned out across campus. When any one of them spotted an attractive stranger of the opposite sex, he or she approached and delivered a sexual proposition, exactly as worded in the scene above.

The results were not surprising. Not a single woman said yes. Frequently they demanded the men leave them alone. By contrast, 75 percent of the guys were happy to oblige. Many queried why they needed to wait until the evening. The few who turned down the offer typically apologized for doing so, giving explanations such as "I'm married." Here, at last, was experimental proof that men are easy.

When the students tried a slightly less forward approach and asked, "Will you come over to my apartment tonight?" the results were almost identical. Sixty-nine percent of the men assented, versus only 6 percent of the women. But the more innocuous question, "Would you go out with me tonight?" produced an affirmative answer from approximately 50 percent of both men and women. To Clark, this was the most surprising finding. He joked that had he known all he had to do was go up to attractive females, ask them out, and half would say yes, his dating years would have been a lot easier.

It took Clark over a decade to get his experiment published. Journal after journal rejected it. For a while he stopped trying. But then, thanks to a chance conversation at a seminar, Elaine Hatfield came to his aid. (We previously met Elaine when her last name was Walster and she was investigating whether women who play hard to get are more desirable.) Hatfield helped Clark revise the paper and signed on as coauthor. Then they began the process of submitting it all over again. Still they met with rejection. One reviewer wrote:

> The study itself is too weird, trivial and frivolous to be interesting. Who cares what the result is to such a silly question, posed in such a stranger-to-stranger way in the middle of the FSU quadrangle? I mean, who cares other than *Redbook*, *Mademoiselle*, *Glamour*, or *Self*—all of which would cream their jeans to get hold of this study. This study lacks redeeming social value.

Finally, just when the authors were about to give up, the *Journal of Psychology & Human Sexuality* accepted it. Clark and Hatfield then had their revenge. The article generated a huge amount of interest, both from the mainstream media and within the academic community. In 2003 the journal *Psychological Inquiry* hailed the study as a "new classic."

The reason for the experiment's continuing popularity is that it dramatically highlights the differing sexual attitudes of men and women. These attitudes appear to be quite stable over time. Clark repeated the experiment in 1982 and 1990 with virtually identical results.

Why do women say no, and men say yes? Clark considered this a sociobiological legacy. Women, he argued, evolved to be more selective about mates because they could only conceive a limited number of children. They needed to be sure about the father. Men, on the other hand, could father an unlimited number of children, so it was a better strategy for them to be always ready and willing. Many critics disagree. They argue either that these attitudes are merely socially learned behavior, or that the women said no because they deemed the invitation too risky. Clark counters that half the women were willing to go on a date with a total stranger. This may indicate that their behavior was motivated less by fear than by a desire to have more time to assess the potential mate.

Whatever the reason for the differing attitudes, the difference itself appears to be real enough (assuming things haven't changed much since 1990). For this reason, your average man should realize that if a beautiful stranger ever does approach him out of the blue on a college campus and invite him to have sex with her, the appropriate response is not "I'd love to," but "Who is conducting this experiment?"

Clark, R. D., & E. Hatfield (1989). "Gender Differences in Receptivity to Sexual Offers." *Journal of Psychology & Human Sexuality* 2 (1): 39–55.

Counting Pubic Hairs

During the mid-1990s six employees at the Alabama Department of Forensic Sciences received an unusual homework assignment. Each was to go home, have sex with his spouse, and, immediately following intercourse, place a swabby towel beneath the buttocks of his partner and thoroughly comb her pubic hair. This wasn't a lesson in postcoital grooming. The point was to collect any fallen hairs on the towel for later examination. The employees were voluntary participants in an experiment to determine the frequency of pubic hair transfer during intercourse.

Forensic scientists had long been trained to search for foreign pubic hairs on victims of sexual assault. Such hairs, if found, can serve as valuable evidence, either implicating or ruling out suspects. But what forensic scientists didn't know was whether it was actually common for pubic hairs to transfer between partners during intercourse. Should they expect to find transferred pubic hairs frequently, or infrequently? It was the kind of question only a strange experiment could answer.

The Alabama researchers collected 110 pubic-hair-bearing towels from the six couples over a period of a few months. They carefully examined all of them and identified a grand total of 334 pubic hairs, as well as seven head hairs, twenty body hairs, and one animal hair. We won't speculate about where the animal hair came from.

Foreign pubic hairs (i.e., hairs transferred from a spouse) were present on only nineteen of the towels. This gave a fairly low transfer rate of 17.3 percent. The transfer of hairs from women to men proved more than twice as common as the transfer of hairs from men to women. No sexual position appeared to cause significantly more hair transfer than any other position.

Given that the hair collection occurred under ideal circumstances, immediately following intercourse, the researchers determined that, in most cases of sexual assault, pubic-hair transfer probably does not occur. Therefore, "Failure to find transferred pubic hairs does not indicate that intercourse did not occur." These results remain the cutting edge (or should we say shedding edge?) of pubic-hair-transfer science.

Exline, D. L., F. P. Smith, & S. G. Drexler (1998). "Frequency of Pubic Hair Transfer During Sexual Intercourse." *Journal of Forensic Sciences* 43 (3): 505–8.

The Penis Imagined as
a Sperm-Shoveling Scoop

The pursuit of knowledge can take researchers to many exotic, out-of-the-way locations—the depths of the ocean, inside the craters of volcanoes, the surface of the moon. In the case of Gordon Gallup, it took him to the Hollywood Exotic Novelties sex store, where he obtained a latex phallus and an artificial vagina. These were strictly for business, not pleasure.

Back at his lab at the State University of New York at Albany, Gallup whipped up some fake semen. The recipe, for those curious, was 7 milliliters of room-temperature water mixed with 7.16 grams of cornstarch and stirred for five minutes. This produced a substance "judged by three sexually experienced males to best approximate the viscosity and texture of human seminal fluid."

Gallup and his team carefully poured the fake semen into the artificial vagina. Then they fully inserted the latex phallus. They repeated this procedure with phalluses of different sizes and semen of varying consistency.

It wasn't sex-ed day at the lab. The point of all this simulated intercourse was to examine the fluid dynamics of sperm inside the vagina. Gallup theorized that the head of the human penis had evolved its distinctive shape to serve as a kind of semen scoop. This morphology, he argued, would have conferred an evolutionary advantage to a man if he had intercourse with a woman shortly after another man. His penis would scoop out the sperm of his rival and replace it with his own sperm.

Gallup's tests confirmed that the penis indeed scoops sperm from the vagina quite effectively. When a penis was fully inserted into the artificial vagina, "semen flowed back under the penis through the frenulum and then collected over the top of the anterior shaft behind the coronal ridge." When pulled out, the penis brought with it as much as 90 percent of the sperm.

Gallup's theory stirred up controversy. Critics pointed out that if the penis does work as a scoop, then continued thrusting after ejaculation would be evolutionarily disadvantageous. The man would simply scoop out his own sperm. Gallup countered by noting the existence of a number of biological mechanisms that inhibit postejacu--latory thrusting, such as penile hypersensitivity, loss of erection, and the refractory period (the postcoital period during which hormones temporarily shut down the male sexual response).

Gallup was no stranger to controversy. In 2002 he had made headlines when he announced the results of a study indicating semen may act as an antidepressant. Of the 293 women who participated in his study, those whose partners *did not* use condoms scored higher, on average, on tests assessing happiness than women whose partners *did* use condoms. Gallup was quick to note that these results should not be taken as a recommendation for abandoning the use of condoms. Contracting a sexual disease, after all, could prove extremely depressing.

Considering Gallup's two studies together, you might get the idea that the penis is rather like an ice-cream scoop. After all, both are scoops that deliver viscous antidepressants. But there is one huge difference: What the ice-cream scoop delivers may make you happy and enlarge your belly, but it won't make you pregnant.

Gallup, G. G., R. L. Burch, M. L. Zappieri, R. A. Parvez, M. L. Stockwell, & J. A. Davis. (2003). "The Human Penis as a Semen Displacement Device." *Evolution and Human Behavior* 24: 277–89.

Mommy Likes Clowns

Couples have been known to do many unusual things to increase their odds of conceiving a child—having sex only in certain positions, timing their lovemaking to the phases of the moon, and sometimes resorting to making the guy wear heated underwear. But what about hiring a clown? Not to be a sperm donor, but to entertain the woman. If the results of a recent study are to be believed, it might be worth a try.

Dr. Shevach Friedler arranged for women undergoing in vitro fertilization embryo-transfer procedures at the Assaf Harofeh Medical Center in Israel to enjoy a "personal encounter with a professional medical clown." The clown performed the same bedside act for all the women. Dressed as a character called Chef Shlomi Algussi, he did magic tricks and told jokes.

Thirty-three of the ninety-three women who received the clown therapy conceived. This was a success rate of 35.5 percent. By contrast, only eighteen of ninety-three women who didn't meet Chef Shlomi conceived—a significantly lower rate of 19 percent. The clown literally worked magic with the patients. Friedler concluded, "Medical clowning has been shown as an original, effective adjunctive intervention having a beneficial effect upon outcome of IVF-ET."

Friedler himself had studied mime in France before becoming a doctor. This gave him the idea for the clown therapy. He suspected humor might relieve some of the stress the women were experiencing, and thus boost their odds of getting pregnant.

Of course, women who suffer from coulrophobia—a fear of clowns and mime artists—should probably avoid the use of medical clowning. Unless, that is, they hope to benefit from its contraceptive possibilities.

Friedler, S., et al. (2006). "The Effect of Medical Clowning on In Vitro Fertilization and Embryo Transfer Treatment," in *Abstracts of the 22nd Annual Meeting of the European Society of Human Reproduction and Embryology*. Poster 563. i216.

CHAPTER SEVEN

Oh, Baby!

Experimenters love babies—partly because babies are cute and smell good, but mostly because babies make fascinating research subjects. They allow experimenters to get a glimpse of the human mind in its original state, before the world has left its mark. So there's no shortage of odd situations infants have been placed in for the sake of science. The experimental appeal of newborns dates all the way back to the seventh century BC when King Psammetichus I ruled Egypt. Psammetichus believed his people were the most ancient in the world, but the Phrygians also claimed this title. To settle the dispute, Psammetichus devised an experiment. He confined two infants to an isolated cottage. Every day a shepherd fed and cared for them, but never uttered a word. Psammetichus reasoned that the first word the children spoke would be the original, natural language of humankind. Two years passed, then one day the shepherd opened the door and heard the children shouting "becos." The shepherd informed the king, who inquired what this meant. He was told "becos" was the Phrygian word for bread. Therefore, Psammetichus yielded the claim of greater antiquity to the Phrygians. However, modern scholars have suggested that—assuming the story is true—the children were possibly mimicking the sounds of the sheep and goats the shepherd tended and the shepherd simply misunderstood their cries. Appropriately, this would make Psammetichus's study not only one of the first experiments in recorded history, but also one of the first examples of experimental error.

Little Albert and the Rat

The Harriet Lane Nursery Home, 1919. An attractive young woman releases a rat onto a mattress. The rat twitches its nose, sniffing the air. Then it scurries across the fabric toward a pudgy, round-faced infant. "Little Albert, look at the rat," the woman says. Albert gurgles and reaches out his hand. His fingers brush the rat's fur. At that instant—BANG!—a middle-aged man standing behind the child smashes a hammer against a steel bar. The sound rings out like a gunshot. Albert flinches with shock. He sucks in his breath, his lips tremble, and he begins to cry.

The man with the hammer was John Broadus Watson, a professor of psychology at Johns Hopkins University. Depending on whom you ask, he was either senselessly scaring a child or conducting an experiment that would revolutionize modern psychology.

The purpose of the experiment was simple. Watson hoped to find out whether he could make eleven-month-old Albert fear a white rat. Why he wanted to do this was a little more complicated.

Let's begin with the experiment itself. When Watson first

met Albert—or Little Albert, as he became popularly known—the young boy didn't fear many things. Watson described him as "stolid and unemotional" and "extremely phlegmatic." When presented with a variety of objects—a white rat, a rabbit, a dog, a monkey, a Santa mask, a burning newspaper—Albert stared at them, showing little reaction.

Watson set out to break down Albert's stoutheartedness and teach him fear. During the first experimental session, Watson's assistant, a graduate student named Rosalie Rayner, showed Albert a rat. Twice Albert reached out to touch it, and each time Watson struck the hammer against the bar. Albert started violently when he heard the jarring sound, but he didn't cry. Not yet.

The experimenters gave Albert a break for a week, then brought him back for more. Again and again they showed Albert the rat and hit the steel bar as soon as he touched the animal. Pretty soon Albert grew wary of the rat. He was learning to associate it with the scary noise. But he didn't easily give in to fear. Instead, he stubbornly stuck his thumb in his mouth and tried to ignore the experimenters. Frustrated, Watson pulled the child's thumb out of his mouth, showed him the rat again, and then—BANG!—hit the bar.

After repeating this process seven times, Watson and Rayner finally achieved the desired result. Albert took one look at the rat and, without the bar having been struck, burst into tears. He had learned to fear the rodent.

Over the next month and a half, Watson and Rayner periodically retested Albert. His fear of the rat not only remained—though they did refresh his memory of the scary noise a few times—but also spread to similar objects he hadn't feared before. The brave little boy had become a coward. He now whimpered and cried when presented with a rabbit, a dog, a fur coat, cotton wool, a Santa mask, and even Watson's hair.

Watson had hoped to reverse the process, removing Albert's newly acquired fears, but he never got the chance. Albert's mother, who worked at the nursery as a wet nurse, left and took her son with her. Nothing is known of what became of the boy.

Watson's fear-reversal technique would have consisted of teaching Albert to associate the rat with pleasurable sensations. Watson wrote that he could have achieved this in a number of ways. For instance, he could have given Albert candy whenever Albert saw the rat, or he could have manually stimulated the child's erogenous zones in the presence of the rat. "We should try first the lips, then the nipples and as a final resort the sex organs," Watson noted. Perhaps it's just as well Albert got out of there when he did.

So what exactly did Watson think he was achieving by teaching an infant to fear a rat? It was all part of his attempt to make psychology less philosophical and more scientific. Psychologists, he felt, spent too much time pondering vague, ambiguous things like emotions, mental states, and the subconscious. He wanted psychologists to focus on measurable, visible behaviors, such as the relationship between stimulus and response. Something happens to a person (a stimulus occurs) and the person responds in a certain way. Action A causes Response B. All very quantifiable and scientific. In Watson's mind, there was no need for patients to lie on a couch and talk about their feelings. Instead, by studying the stimulus-response interaction, scientists could learn how to control human behavior. It was just a matter of applying the right stimulus to trigger the desired response. He once famously boasted:

Give me a dozen healthy infants, well-formed, and my own specified world to bring them up in and I'll guarantee to take any one at random and train him to become any type of specialist I might select—doctor, lawyer, artist, merchant-chief and, yes, even beggar-man and thief, regardless of his talents, penchants, tendencies, abilities, vocations, and race of his ancestors.

Watson designed the Little Albert study to prove that a simple stimulus, such as banging on a steel bar, could produce a wide range of complex emotions in a child—namely, fear of rats, dogs, rabbits, wool, hair, fur coats, and Santa Claus. The experiment was a deliberate swipe at Freudian psychology, which, Watson sneered, would probably have attributed Albert's fears to repressed sexual urges.

Watson made his case well, and behaviorism, as he named his approach, became a dominant school in psychological research for the next fifty years. Which is why many would call the Little Albert experiment revolutionary. Many others, however, argue that while the experiment may have been good drama, it was bad science and didn't prove anything, except that any child will cry if you harass him enough.

Watson would have liked to continue his infant studies, but he never had the chance. His wife smelled a rat and found out his affair with his graduate student assistant, Rosalie Rayner. The judge in the subsequent divorce proceedings remarked that the doctor was apparently an expert in *mis*-behavior. Because of the scandal, Watson was forced to leave Johns Hopkins.

Lurid rumors would later suggest Watson was not only sleeping with Rayner, but also using her as a subject in various sex experiments, measuring physiologic responses such as her pulse rate as he made love to her. This is offered as the true reason for Watson's dismissal—the story being that his wife discovered his records of this research. However, there is no good evidence to substantiate such gossip. Watson frequently did express an interest in studying the human sexual response, but if he had conducted such experiments, he probably would have mentioned them to someone. After all, he wasn't one to shy away from the frank discussion of sexuality.

Blacklisted by academia, Watson headed to Madison Avenue and the lucrative world of advertising. There he put his stimulus-response theories to great effect, introducing techniques that are used to this day. He designed successful ad campaigns for coffee, baby powder, and toothpaste, among other items. Watson figured that getting a consumer to perform a desired action, such as buying a product, was simply a matter of applying the correct stimulus. One stimulus that invariably worked was sex. If Watson could have reached out and directly stimulated consumers' erogenous zones, he would have. Instead he had to settle for visual arousal. So the next time you see bikini babes selling beer, know that you have John Watson to thank.

Watson, J. B., & R. Rayner (1920). "Conditioned Emotional Responses." *Journal of Experimental Psychology* 3 (1): 1–14.

Self-Selection of Diet by Infants

"Eat your vegetables."
"I don't want to."
"You're not leaving this table until you finish them."
"Waaaahhhhhhh!"

Scenes like this are all too common at dinner tables, as desperate parents try to force good nutrition on their resistant kids.

Wouldn't it be easier just to let children eat whatever they want? Whenever frazzled parents compare notes, someone inevitably makes this tempting suggestion. And someone else is sure to chime in with, "Yeah, wasn't there a doctor who conducted an experiment that proved that if kids are allowed to eat whatever they want, they naturally choose a well-balanced diet?"

Yes, there was such a doctor. Her name was Dr. Clara Davis. But despite the urban-legend-style rumors that circulate about her study, what, if anything, it proved is up for debate.

Davis's study, conducted in 1928, was a culinary version of Psammetichus's language experiment. Psammetichus had hoped to discover the natural language of humans by observing children who had never heard others speak. Similarly, Davis hoped to discover humankind's natural diet by observing children who had never been fed solid food before and were therefore free of adult tastes and habits. Would they prefer a carnivorous, vegetarian, or omnivorous diet? And more important, would their bodies automatically make them desire the foods that met their nutritional needs, providing them with a well-balanced diet?

Davis used as her subjects three newly weaned infants between seven and nine months old at Cleveland's Mount Sinai Hospital. She arranged for the infants—Donald, Earl, and Abraham—to eat alone, away from other children. At the beginning of each meal, a nurse placed a tray in front of the boy. On the tray were dishes containing different foods —chicken, beef, cauliflower, eggs, apples, bananas, carrots, oatmeal, and the like. The child was free to eat whatever he wanted from this tray, in whatever quantity he desired. The nurses had specific instructions about how the feeding should occur:

> Food was not offered to the infant either directly or by suggestion. The nurses [sic] orders were to sit quietly by, spoon in hand, and make no motion. When, and only when, the infant reached for or pointed to a dish might she take up a spoonful and, if he opened his mouth for it, put it in. She might not comment on what he took or did not take, point to or in any way attract his attention to any food, or refuse him any for which he reached. He might eat with his fingers or in any way he could without comment on or correction of his manners. The tray was to be taken away when he had definitely stopped eating, which was usually after from twenty to twenty-five minutes.

Initially the children didn't display great manners. They thrust their whole hand or face into the dishes. They threw food on the floor. When they tasted food they didn't like, they spluttered and spat it out. But soon they figured out the routine. Like little princes, they would point at a dish with their stubby fingers, open their mouths expectantly, and wait for the food to arrive.

Two of the children stayed on the diet for six months, the third for a year. At the end of this period, Davis examined her data and tried to draw some conclusions.

She found it impossible to discern any innate dietary preferences beyond noting that humans are definitely omnivorous. At first the children sampled dishes randomly, but soon they developed favorites they sought out no matter where the dishes were placed on the tray. However, their favorites changed unpredictably every few weeks. The nurses would say, "Donald is on an egg jag this week." Or perhaps it would be a "meat jag" or a "cereal jag." Milks, fruits, and cereals were, by volume, the foods the kids chose most often. They chose bone product, glandular organs, and sea fish least often.

Did the children make sensible choices that provided for their dietary needs? Here we should note that Davis conducted her experiment before scientists had a clear understanding of the role vitamins play in our bodies' health. So from a modern perspective, her analysis seems less than rigorous. She basically eyeballed the kids, decided they looked plump and well nourished, and declared they had done a fine job of managing their dietary needs. She noted the children had come down with a series of illnesses during the course of the experiment—including influenza, whooping cough, and chicken pox—but she didn't regard this as significant. And perhaps it wasn't, given the germs they were exposed to at the hospital.

However, Davis did offer one tantalizing piece of evidence to indicate the existence of a self-regulating dietary mechanism. One of the children, Earl, had begun the experiment with a case of rickets. Davis added a dish of cod liver oil to his tray in the hope that he would voluntarily down the fishy liquid. Surprisingly, he did—for three months, until his rickets were cured. Then he stopped eating it. Perhaps his body made him desire the medicine he needed. Or perhaps it was random chance. It's difficult to say.

Davis declared her experiment a success, but she readily admitted this wasn't an invitation to laissez-faire rules in the dining room. As her critics often point out, and as she acknowledged, there was a trick to her experiment: The children had no unhealthy options. Davis gave the infants no canned, dried, or processed foods, no peanut butter sandwiches, no chocolate milk, no cheese, no butterscotch pudding, no ice-cream sundaes—in other words, no tasty but unhealthy enticements to lure them from the path of righteous eating. Davis had stacked the deck in her favor. As long as the kids consumed enough food, the odds were they would get a balanced diet.

So if you're a parent and want to try your kids on Davis's eat whatever-you-want diet, feel free. It probably won't do any harm. But realize the first step is to eliminate junk food. All Happy Meals, crisps, pizza, and fizzy drinks must go. Then watch your child's eyes widen with delight as you offer him a selection of cooked marrow, spinach, raw carrots, unprocessed whole wheat, and cauliflower. You probably won't be able to count to five before you hear the *"Waaaahhhhhh!"* Most parents will quickly conclude it's easier to stick with the pizza and force the little darlings to eat a few vegetables now and then.

Davis, C. M. (1928). "Self-Selection of Diet by Newly Weaned Infants: An Experimental Study." *American Journal of Diseases of Children* 36 (4): 651–79.

The Masked Tickler

The Leuba household, 1933. An infant lies awake in a crib. Suddenly a man opens the bedroom door and walks toward the baby. He is wearing a cardboard mask with narrow slits cut out for his eyes. He stands over the child without saying a word. He holds himself tense, as though consciously trying to suppress his body language. Then he reaches down and pokes the child gently beneath the armpit. The child looks up and smiles. The man nods slightly, his hands continuing to move as if following a predetermined pattern. He pokes along the ribs, under the chin, on the side of the neck, inside the knees, and finally runs a finger along the soles of the feet, all without making a sound. When the child laughs the man abruptly steps away, picks up a journal, and busies himself writing in it for the next few minutes.

A stranger observing these events might have been alarmed. What was the masked man doing? Was he an intruder? Did he mean to kidnap the child? But there was no cause for concern. The man was Dr. Clarence Leuba, the child's father and a professor of psychology at Antioch College. His mysterious actions were part of an experiment to understand the phenomenon of tickling.

Leuba wondered why people laugh when they're tickled. Tickling, after all, is not self-evidently funny. Many find it painful, especially if done to excess. So is the laughter a learned response, something we pick up as infants by observing others laugh when tickled, or is it an innate response?

Leuba reasoned that if laughter is a learned response to tickling, then it should be possible to raise a child who would not laugh when tickled. For this experiment to work, the child would need to be shielded from all displays of tickle-induced laughter. He could not hear his mother chuckle as she bent down to tickle him. Nor could he observe siblings cackle with glee as fingers sought out their armpits. If the laughter response was innate, the child would eventually laugh anyway, but if the response was learned, the child would respond to tickling with a blank stare.

Such an experiment would not be easy to conduct, but for the sake of science, Leuba decided to give it a try.

He volunteered his own household as the experimental setting. He could scarcely invade another family's home to monitor their tickling behavior around the clock. And he would use his own newborn son as the test subject. In his research notes he designated the boy as R. L. Male.

Only one serious obstacle lay in Leuba's path—his wife. With a single giggle she could ruin everything. In his report he cryptically noted that "the mother's cooperation was elicited." This doesn't give us much of a clue about her reaction to the idea. Did she laugh, or roll her eyes and say, "For God's sake, Clarence, why?" We will never know. But somehow Leuba did obtain her promise of assistance.

And so the Leuba household became a tickle-free zone, except during experimental sessions when R. L. Male. was subjected to laughter-free tickling. During these sessions, strict guidelines were followed. The tickler concealed his face behind a twelve-by-fifteen-inch piece of cardboard. He maintained a "smileless, sober expression, even though his face was hidden behind the cardboard shield." And the tickling followed a regular pattern—first light, then vigorous—in order of armpits, ribs, chin, neck, knees, then feet. Occasionally Leuba used a tassel.

Even outside of the experimental sessions there were guidelines to be followed: "The baby should never be tickled when he can see or hear a person laughing or when laughter is being produced or facilitated in him by jouncing, a peek-a-boo game or the like." Perhaps, for the mother's benefit, this rule was posted on the refrigerator.

The experiment proceeded. The rules were enforced. But, unfortunately, a few lapses occurred. On April 23, 1933, Leuba recorded a confession from his wife—on occasion, after her son's bath, she had "jounced him up and down while laughing and saying 'Bouncy, Bouncy.' " Perhaps this was enough to ruin the experiment. That is not clear. What is clear is that by month seven, R. L. Male was happily screaming with laughter when tickled.

Leuba was undeterred. When his daughter, E. L. Female, was born in February 1936, he repeated the experiment, with the same result—laughter at seven months. You sense a certain regret in Leuba that his children kept turning out so normal. Laughter, Leuba concluded, must be an innate response to being tickled. There went his chance at raising the world's first untickleable child.

Did the children suffer any psychological damage as a result of the experiment? Did they develop a profound fear of masked men, a fear they could never quite understand? We don't know. A follow-up study was not conducted.

Leuba's study received scant attention from the scientific community, although fifty-eight years later Dr. Harris did cite it in her mock-tickle-machine study (see chapter two). However, a faint echo of Leuba's experiment can be found in American popular culture. In the 1970s a masked villain named the Tickler appeared on an episode of the TV show *Spidey Super Stories*. With feathers attached to his fingers, the Tickler incapacitated his victims with laughter before robbing them. A mere coincidence? Probably. But it's fun to imagine Dr. Leuba, frustrated by the slow progress of his research, sneaking out of his home one evening and embarking on a new career as a comic-strip super villain.

—————

Leuba, C. (1941). "Tickling and Laughter: Two Genetic Studies." *Journal of Genetic Psychology* 58: 201–9.

A Girl Named Gua

From the start, Winthrop and Luella Kellogg knew their adopted daughter, Gua, was different. It wasn't just her physical appearance—her lumpy overhanging brows or the black hair that hung down on either side of her face like sideburns. There were other things. Her jumping, for instance. Her great strength would send her flying through the air, from window to bed or from porch to ground. When startled she would instinctively vault forward two or three feet. All the other babies would simply look around bewildered.

"Of course she's different," people told them. "She's a chimp!" But the Kelloggs were determined to ignore this surface difference. In their eyes, she was just a little girl—though admittedly one hairier than most.

The experiment occurred to Winthrop Kellogg in 1927, while he was a graduate student in psychology at Columbia University. He had read an article about cases of wild children raised by animals. Even after being returned to society, such children often continue to act more animal than human. They grunt and crawl around on all fours. Kellogg wondered what would happen if the situation were reversed. If an ape were raised by a human family, would it learn to act like a human, walking upright and eating with a knife and fork? Kellogg suggested to his wife, Luella, that they take a chimpanzee baby into their home and find out. They would never put it in a cage or treat it like a pet. Instead they would cuddle it, talk to it, spoon-feed it, and clothe it like a child. At the least, Kellogg figured such an experiment would offer valuable insights into the relationship between environment and heredity.

Luella resisted the idea. Evidently she was the sensible one. But in 1931 a grant from the Social Science Research Council gave Winthrop enough money to conduct the experiment, and the Yale Anthropoid Experiment Station in Orange Park, Florida, offered him a chimp. Grudgingly, Luella agreed to go along with her husband's plan, so they packed their bags and moved to Florida to meet their new daughter.

One other factor made the timing perfect. The Kelloggs had recently had a child of their own, a son named Donald. This presented Winthrop a unique opportunity to raise a chimp and a human side by side, allowing him to collect detailed data about the comparative rates of development of the two species.

Gua arrived at the Kellogg household on June 26, 1931. She was seven and a half months old. Donald was ten months old.

During the first meeting of the two infants, the parents hovered over them nervously, ready to intervene at the first sign of tension. But there was no need. Donald was immediately fascinated by Gua. He reached over and touched her. Initially Gua showed little corresponding interest, but by the next time they met, she had warmed up to him considerably. She leaned over and kissed him. From that moment on, the two were inseparable.

The experiment proved to be a full-time job. Not only were there the usual tasks involved with caring for infants —bathing, feeding, changing diapers—but the Kelloggs also kept themselves busy recording details of how the babies ate, slept, walked, and played. They noted unusual emotional reactions. For instance, Gua had an unaccountable fear of toadstools. They wrote down responses to smells—Donald liked perfume but Gua hated it. They even recorded what sound a spoon made when it was knocked against the infants' heads:

> The differences between the skulls can be audibly detected by tapping them with the bowl of a spoon or with some similar object. The sound made by Donald's head during the early months is somewhat in the nature of a dull thud, while that obtained from Gua's is harsher, like the crack of a mallet upon a wooden croquet or bowling ball.

The Kelloggs also devised tests to measure Donald's and Gua's abilities. For instance, the suspended-cookie test—how quickly could the infants figure out how to reach a cookie suspended by a string in the middle of the room? And the sound-localization test—with hoods over their heads, could they locate where a person calling them by name was standing? Gua reliably performed better on these tests than Donald, demonstrating that chimps mature faster than humans. So score one for the chimp.

But the Kelloggs were interested not only in Gua's development, but also in how humanized she was becoming. Here the results were mixed. Gua picked up some human behaviors. She often walked upright, and she ate with a spoon. But in other ways she remained decidedly chimplike. She was, in the words of the Kelloggs, a creature of "violent appetites and emotions." Simple things, such as people having changed their clothes, would confuse and frighten her. The ability to speak eluded her, despite Winthrop's repeated efforts to make her say "Papa." And she failed entirely to grasp the concept of pat-a-cake—a game that Donald understood right away. So score one for the human!

To be fair, Donald wasn't proving to be much of a speaker, either. Nine months into the experiment, he had only mastered three words. Which left pat-a-cake as the sole arena in which he truly reigned victorious over the chimp. But what he did say began to worry the Kelloggs. One day, to indicate he was hungry, he imitated Gua's "food bark." Suddenly, visions of their son transforming into a wild child, grunting and crawling on all fours, danced before their eyes. Perhaps, the Kelloggs realized, some playmates of his own species would be better for his development. So on March 28, 1932, they shipped Gua back to the primate center. She was never heard from again.

Could Gua have been humanized had the experiment continued longer than nine months? The answer is certainly no. Primatologists now know enough about chimpanzees to state this definitively. Chimps are wild animals. Their inherent wildness eventually reasserts itself, even if they're raised in a human family. So it's just as well the Kelloggs ended the experiment when they did.

Unfortunately, every year people insist on learning this lesson the hard way. It's become quite popular to purchase baby chimps as pets—even though they can cost more than forty thousand dollars. But a few years later, owners have on their hands a full-grown, enormously strong animal that requires skill and training to handle. Mature chimps can be willful, mischievous, and destructive. If they're bored they look for trouble. They pull down drapes and knock over furniture. They're smart enough to know the one thing a person values most in the house, and they may purposefully decide to smash it. Come to think of it, they're not that different from a typical younger sibling. So perhaps Gua could have been a normal little sister for Donald after all.

———

Kellogg, W. N., & L. A. Kellogg (1933). *The Ape and the Child: A Study of Environmental Influence upon Early Behavior.* New York: McGraw-Hill Book Company, Inc.

Baby in a Box

The first baby was a lot of work. There was all that laundry and cleaning, and if he wasn't careful when he bent down to lift the child out of her crib, he risked spraining his back. So in 1943, when Burrhus Frederic Skinner's wife became pregnant for a second time, he decided to use his scientific training to reduce the drudgery of baby care. He came up with a device he called the mechanical baby tender. It became more widely known as the "baby box."

Skinner's psychological research had well equipped him for gadget making. Over ten years earlier, while a graduate student at Harvard, he had invented a device called an operant chamber, or Skinner Box. The box held an animal, such as a rat or a pigeon; when the animal pressed a lever, it received a reward, usually food. Skinner, an outspoken proponent of behaviorism—the school of psychology pioneered by John Watson of Little Albert fame—used this box to demonstrate that by varying the frequency of rewards, he could dramatically alter the behavior of animals, training them to do just about anything. For instance, a rat named Pliny learned that in order to make its food appear, it first had to pull a lever to make a marble drop from a chute, then pick up the glass ball and place it down a slot.

During World War II Skinner embarked on an even more ambitious project—training pigeons to guide missiles. Strapped into the nose cone, the bird would guide the bomb by pecking at a target on a screen. The weird thing was, the system actually worked—at least as well as any electronic guidance system of the time. But the idea proved too bizarre for the military, which cut funding for the project. Disheartened, Skinner focused his creative energies on building the mechanical baby tender. Compared to a pigeon-guided missile, the new project must have seemed like child's play.

The baby tender was essentially a large box six feet high and two-and-a-half feet wide. The baby sat in a shallow pan about three feet off the ground, peering out at the world through a large safety-glass window that could be slid up and down. A heater, humidifier, and air filter circulated warm, fresh air within the chamber. Insulated walls muffled the noise of the outside world.

The unit offered many conveniences and safety advantages. The heated interior meant the baby didn't need clothes or blankets, just a diaper. So there was less laundry. The window both protected the baby from germs and prevented her from falling out. The mattress consisted of a ten-yard-long sheet of canvas attached to rollers. When it got dirty, the parents simply rolled out a clean section. And because the device was quite tall, parents could place a baby in it without damage to their backs. All in all, the invention was very practical.

Skinner's daughter, Deborah, became the guinea pig on whom he tested the baby tender. After nine months, she was healthy and happy. Skinner, judging his invention to be a success, decided to let the world know about it. Eschewing academic journals, he sent an article to the popular women's magazine *Ladies' Home Journal*. The editors of *Ladies' Home Journal*, recognizing an entertaining oddity when they saw it, published the article—with one slight alteration. They changed Skinner's title from "Baby Care Can Be Modernized" to "Baby in a Box."

Skinner always blamed this one editorial change for the public reaction that followed. No matter how much he later tried to convince people of the benefits of his invention—how much time it would save the mother and how much more comfortable it would make the baby—their reaction consistently remained the same: "You've put a baby in a box!" One angry reader wrote in to a local paper saying, "It is the most ridiculous, crazy invention ever heard of. Caging this baby up like an animal, just to relieve the Mother of a little more work." An entire high school English class wrote directly to Skinner to inform him that "by creating this 'revolutionary product,' you have shown that you are ready to inaugurate a society composed of box-raised vegetables similar to the *Brave New World* of Aldous Huxley." Another critic charitably compared the baby tender to a quick-freeze display case.

The idea that the baby tender was some kind of giant Skinner Box designed to behaviorally condition babies took root. Skinner conceded that, given the similarity between the terms Baby Box and Skinner Box, "it was natural to suppose that we were experimenting on our daughter as if she were a rat or pigeon." But this was not the case. In fact, Deborah's time in the baby tender was more of a trial run than an experiment. Skinner did hope to conduct a formal experiment in which he would compare ten babies raised in the baby tender to ten babies raised in normal cribs, but this study never happened.

The perception of Deborah as the unwitting subject of a

human Skinner Box experiment inspired a series of urban legends that surfaced during the 1950s and '60s. According to these rumors, as an adult Deborah became psychotic, sued her father, and committed suicide. In reality, Deborah grew up quite normal and became a successful London-based artist. Though intriguingly, as art critics have noted, her paintings "appear to represent visions seen through 'glass prisms'— perhaps reflections reminiscent of infant window views."

Not all reactions to the baby tender were negative. A small community of enthusiasts embraced the concept. But in the words of one General Mills engineer whom Skinner approached about producing a commercial version of the device, these supporters tended to be "long-haired people and cold-hearted scientists." This wasn't a demographic General Mills was interested in selling to.

Skinner eventually worked out a manufacturing deal with a Cleveland businessman, J. Weston Judd, who had the inspired idea of marketing the baby tenders as "Heir Conditioners." But Judd turned out to be a con artist who failed to deliver any product and then skipped town with five hundred dollars Skinner had loaned him. In the 1950s an engineer, John Gray, next took up the thankless job of selling baby tenders. He came up with a better name—Air Crib—and actually sold a few hundred units. But when Gray died in 1967, the Air Crib industry died with him. Unless you strike it lucky on eBay, you'll be hard-pressed to get your hands on an Air Crib today.

Ultimately the baby tender was a decent (or, at least, harmless) idea that suffered from a serious image problem. As proof of the basic soundness of the concept, Skinner's advocates point to the hundreds of healthy, sane people who were raised in the devices. But the public remained uncomfortable with the notion of enclosing a child in a box. Perhaps people simply were concerned that, in such a highly engineered environment, a kid would grow up too square.

———

Skinner, B. F. (October 1945). "Baby in a Box." *Ladies' Home Journal*: 30–31, 135–136, 138.

The New Mother

"If you continue to be naughty I shall have to go away, and leave you and send home a new mother with glass eyes and a wooden tail."

So an exasperated mother warns her children in Lucy Clifford's short story "The New Mother," first published in 1882. As Clifford was a writer of dark, gothic fairy tales, her readers knew what to expect. The children keep misbehaving, until the mother sadly packs her bags and departs. Hours later the new mother arrives, announcing her presence with a terrible knocking on the door. The frightened children peer out the window and see her long bony arms and the flashing of her two glass eyes. Then with a blow of her wooden tail, the new mother smashes down the door. Shrieking with terror, the children flee into the forest, where they spend the rest of their lives sleeping on the ground among dead leaves and feeding on wild blackberries, never to return home. That's what they get for being naughty.

Clifford's story sends chills down readers' spines today, over a century after it was written, because it taps into such primal emotions. It takes the image of the mother—the ultimate symbol of love and security—and transforms it into a mechanical terror. The same juxtaposition is what made Harry Harlow's cloth-mother experiments such a sensation when he conducted them in the 1950s, and why they continue to fascinate the public today.

Harlow was a professor of psychology at the University of Wisconsin. He was interested in the nature of love, specifically the love of an infant for its mother. The prevailing psychological wisdom was that love was an overrated and certainly unscientific concept. Psychologists dismissively explained that an infant wanted to be close to its mother simply because she provided milk. That's all love was—an effort to reduce hunger pangs. The same John Watson who terrified Little Albert even warned parents that too much cuddling could warp children's characters, making them whiny and fearful.

Harlow thought this was hogwash. He was sure love was about more than hunger. While raising infant rhesus monkeys at his lab, Harlow had noticed that the tiny primates craved—and seemed to draw strength from—physical contact with their mothers. If separated from their mothers, they would bond with substitutes, lovingly embracing the soft cloth rags used to line the bottom of their cages, in the same way human children become attached to cuddly toy animals and dolls. It seemed to be a drive as strong as hunger.

Harlow decided to test the claim that love is just a desire for milk. He separated infant monkeys from their mothers at birth and put them in a cage with two surrogate mothers of his own design. He called the first surrogate "cloth mother." She was a block of wood wrapped in rubber, sponge, and terry cloth and warmed by a lightbulb. She was, Harlow enthused, "a mother, soft, warm, and tender, a mother with infinite patience, a mother available twenty-four hours a day, a mother that never scolded her infant and never struck or bit her baby in anger." However, she provided no milk. All she could offer was a warm, soft surface to cuddle against.

The second surrogate was a "wire-mesh mother." Her steel-wire frame wasn't cuddly at all, but she did supply milk.

Would the infants go for cuddles or for food? Harlow carefully recorded the amount of time the babies spent with each mother, but it soon became apparent that, in the eyes of the monkeys, there was no doubt which mother was better. They spent almost all their time snuggling with cloth mother, only suckling at wire-mesh mother's teat for a few brief seconds before frantically running back to the security of cloth mother. Clearly, these babies cared more about cuddling than about nourishment.

This experiment demolished in one fell swoop decades of psychological dogma. But Harlow wasn't finished.

Although the infants clung desperately to cloth mother, she clearly wasn't a great parent. Her babies grew up strange—timid and antisocial. They cowered in corners and shrieked as people walked by. Other monkeys shunned them. Wire-mesh-mothered monkeys fared even worse. Harlow realized his surrogate mothers still lacked essential features. He set out to determine scientifically what these might be. What were the significant variables in the relationship between a child and its mother?

He began with texture. He wrapped his surrogate mothers in different materials—terry cloth, rayon, vinyl (which he called the "linoleum lover"), and rough-grade sandpaper. The infants definitely preferred the terry cloth mother and showed more self-confidence in her presence. So Harlow concluded that a good mother must be soft.

Next he investigated temperature. He created "hot mamma" and "cold mamma." Hot mamma had heated coils in her body that raised her temperature. Chilled tubes of water ran through cold mamma. As far as the monkeys were concerned, cold mamma might as well have been dead. They avoided her at all costs. Conclusion—a good mother must be warm.

Finally, Harlow examined motion. Real mothers are always walking around or swinging from trees. To simulate this, he came up with "swinging mom." Swinging mom, hung from a frame like a punching bag, dangled two inches off the floor. Harlow quipped, "There is nothing original in this day and age about a swinger becoming a mother, and the only new angle, if any, is a mother becoming a swinger." Surprisingly, the monkeys loved swinging mom best of all. And under her care, they grew up to be remarkably well adjusted— or as well adjusted as could be expected for a child who has a swinging cloth bag for a mother. So the final tally was that good mothers must be soft, be warm, and *move*.

William Mason, who worked for a while in Harlow's lab after obtaining his Ph.D. from Stanford, later extended this work and came up with the perfect surrogate mother for a baby monkey. She fit all the criteria. She was soft and warm, and moved. She also happened to be a mongrel dog. Mason's dog-raised monkeys turned out strikingly normal. They were bright, alert, and happy little creatures, though perhaps slightly confused about their identity. Remarkably, the dogs didn't seem to mind the little monkeys hanging off them.

Harlow's work then took a darker turn. Having determined the qualities that fortify the love between a baby and its mother, he set out to discover whether these bonds of love could easily be broken. He wanted to create a parallel to human children who experience poor parenting, to help unravel some of the resultant problems. So he created mothers that inflicted various forms of abuse on the babies. There was shaking mom (who at times shook so hard she flung her infant across the room), air-blast mom (who occasionally blasted her babies with violent jets of compressed air), and brass-spike mom (from whom blunt brass spikes periodically emerged). Whatever cruelties these mothers dealt out, their babies would simply pick themselves up and crawl back for more. All was forgiven. Their love could not be shaken, dented, or air-blasted away. Very few infant monkeys were involved in these tests, but the results were clear.

It's worth noting that although Harlow did identify some of the qualities a competent mother should have, his surrogates reliably failed the ultimate test. An infant monkey would never, ever have chosen one of them over a living, breathing female of its own species. Which shows that, when it comes to motherly love, there is no substitute for the real thing.

_____

Harlow, H. (1958). "The Nature of Love." *American Psychologist* 13 (12): 673–85.

Braking for Baby

The squad car speeds down the street, sirens wailing, in hot pursuit of a criminal. It rounds a corner, and suddenly, just yards ahead, a person is crossing the street and—OH MY GOD!—she's pushing a baby stroller! The cop jams on the brakes and twists the wheel sideways. Tires squeal. Rubber skids against the tarmac. The car veers up onto the curb, hits a fire hydrant, and slams into the side of a building. Water sprays everywhere. But the stroller and its occupant are untouched. Inside the car, the cops breathe a sigh of relief. The criminal, however, is long gone.

Hollywood car chase scenes have taught us that people will go to great lengths to avoid hitting a baby stroller—far greater lengths, it seems, than they'll go to avoid running over an adult. Is this true? Are drivers in real life more careful to avoid baby strollers than they are to avoid grown-ups?

In 1978 researchers at UCLA put this idea to an experimental test. At a busy four-lane street in Los Angeles, a young female experimenter stepped out into the crosswalk and waited for cars to stop so she could proceed across. She either stepped out alone, pushing a shopping cart, or pushing a baby stroller. An observer counted the number of cars that passed in each situation before a driver stopped for her.

Thankfully, the researchers didn't put a real baby in danger. The stroller was empty, and drivers could easily see this. But the researchers believed this wouldn't matter. They hypothesized that the mere presence of a stroller would encourage drivers to stop more readily.

They were right. When the experimenter stood alone, an average of almost five cars drove by before one stopped. In the shopping-cart condition, the average was three cars. But with a baby stroller in front of her, the number of passing cars dropped to one.

The researchers theorized that drivers stopped sooner for the stroller because babies—and, by extension, all things baby related—act as anger inhibitors. They trigger nonviolent, courteous impulses in people. Strong taboos against harming small children exist in all cultures. Even monkeys share this sentiment. If male monkeys want to avoid being attacked, they often sidle up to infants.

Intriguingly, the researchers stumbled upon a parallel phenomenon during the course of their experiment. They noticed that when the woman was standing alone, certain types of drivers stopped far more readily than others:

> Young male drivers, in particular, seemed more inclined to stop for the attractive female experimenter when she was without the baby stroller (and in a few instances to actually try and converse with her). It would be desirable in future research to replicate the present findings with other pedestrians such as older men.

So perhaps we should add a corollary to the observation that people stop more readily for babies. They also stop frequently for babes.

———

Malamuth, N. M., E. Shayne, & B. Pogue (1978). "Infant Cues and Stopping at the Crosswalk." *Personality and Social Psychology Bulletin* 4 (2): 334–36.

The Ultimate Baby Movie

You see them at playgrounds, chasing after children, camera in hand. Or at restaurants, filming away as an infant hurls food on the floor. They're proud parents, determined to preserve for posterity every moment of their kid's childhood—his first step, his first bite of carrots, his first inarticulate gurgle. Later these archived memories will be sprung on guests who thought they had been invited over for a no-strings-attached dinner.

Every year proud parents collectively shoot hundreds of thousands—perhaps millions—of hours of baby movies. But no one shoots more than Deb Roy.

As of January 2006, when his kid was six months old, Roy

had amassed 24,000 hours of baby video. By the time his child is three, Roy hopes to have 400,000 hours of audio and video, enough to make any dinner guest shudder.

Roy has managed this feat by installing eleven overhead omnidirectional megapixel fish-eye video cameras in the rooms of his home, as well as fourteen microphones. Literally every move and utterance his baby makes is recorded. The data, almost three hundred gigabytes' worth of it a day, is continuously streamed to a five-terabyte disk cache in his basement. Since installing the system, he's seen his electricity bill quadruple.

Roy isn't doing this for the sake of parental pride, though that certainly plays a part. He is head of the MIT Media Lab's Cognitive Machines Research Group. When his wife became pregnant he realized he had the perfect opportunity to study how children learn language. Roy plans on recording almost everything his son hears and sees from birth until the age of three (in mid-2008). Powerful computers at the MIT media lab will then analyze the footage, searching for the visual and verbal clues the child has used to construct his understanding of language. A one-million-gigabyte storage system—one of the largest storage arrays in the world—has been built to hold all the data. The MIT engineers will then attempt to build a model of language acquisition out of the data. With luck, this model can be used to design a machine learning-system that can mimic a human baby's ability to learn language.

Roy calls his experiment the Human Speechome Project. Speechome stands for "Speech at home." The media, however, have dubbed it the Baby Brother Project. But, unlike the *Big Brother* contestants, Roy hasn't totally sacrificed his privacy. The cameras in every room have an ON/OFF switch, as well as an OOPS button that deletes the last few minutes of activity. Roy notes that the oops button was used 109 times during the first six months the system was in place, although he doesn't state why. If it was used because, "Oops, I just said a bad word," that omission could undermine the purpose of the project. MIT analysts will be scratching their heads wondering, "How in the world did he learn to say that?" Of course, Roy can always do what millions of parents do—blame it on the TV.

———

Roy, D., et al. (2006). The Human Speechome Project. Presented at the 28th Annual Conference of the Cognitive Science Society. Available online: http://www.media.mit.edu/press/speechome/speechome-cogsci.pdf.

CHAPTER EIGHT

Toilet Reading

Back in the Middle Ages, toilets—the few that existed—were placed at the top of castle turrets. Waste products slid down a chute into the moat below. This was one reason swimming the moat was an unappealing prospect. The anonymous author of *The Life of St. Gregory* praised the solitude of this lofty perch for the "uninterrupted reading" it allowed. This admission made him one of the first bathroom readers in recorded history. Today many people do much of their best reading while on the loo. It is possible you are reading these very words in such a situation. Toilet readers can be divided into two types: relaxers and workers. For relaxers the toilet is a place of retreat and tranquillity. Dr. Harold Aaron, author of *Our Common Ailment, Constipation,* notes, "Reading gives the initial feeling of relaxation so useful for proper performance." Workers, on the other hand, dare not waste even those few minutes spent attending to bodily functions. Lord Chesterfield tells of "a gentleman who was so good a manager of his time that he would not even lose that small portion of it which the call of nature obliged him to pass in the necessary-house; but gradually went through all the Latin poets, in those moments." This chapter is dedicated to toilet readers of all persuasions. Gathered in the following pages are unusual experiments that speak, in some way, to loo-related themes. May they help set the mood for peak performance.

The Doctor Who Drank Vomit

The yellow-fever patient groaned as he lay in bed. His skin had a sickly lemon tinge, marred by red and brown spots. The smell of decay clung to him. Suddenly he jerked upward and leaned over the side of the bed. Black vomit, like thick coffee grounds, gushed from his mouth. A young doctor sitting by his side expertly caught the spew in a bucket and patted the patient on the back. "Get it all out," he said. A few final mucus-laced black globs dribbled from the patient's mouth before the man collapsed onto the bed. The doctor swirled the steaming liquid in the bucket a few times, examining it closely. The stench of it was overpowering, but barely a flicker of disapproval registered on the doctor's face. Instead he calmly poured the vomit into a cup, lifted it to his lips, and slowly and deliberately drank it down.

The vomit-imbibing doctor was Stubbins Ffirth. As successive yellow-fever epidemics devastated the population of Philadelphia during the early nineteenth century, Ffirth made a name for himself by courageously exposing himself to the disease to prove his firm belief that yellow fever was noncontagious.

Ffirth confessed that when he first saw the ravages of yellow fever he, like everyone else, believed it to be contagious. But subsequent observation dissuaded him of this.

The disease ran riot during the sweltering summer months, but disappeared as winter approached. Why, he wondered, would weather affect a contagious disease? And why didn't he grow sick, despite his constant contact with patients? He concluded yellow fever was actually "a disease of increased excitement" brought on by an excess of stimulants such as heat, food, and noise. If only people would calm down, he theorized, they would not develop the disease.

To prove his noncontagion hypothesis, Ffirth devised a series of tests. First he confined a dog to a room and fed it bread soaked in the characteristic black vomit regurgitated by yellow-fever victims. (The blackness is caused by blood hemorrhaging from the gastrointestinal tract.) The animal did not grow sick. In fact, "at the expiration of three days he became so fond of it, that he would eat the ejected matter without bread." Pet-food manufacturers might want to take note.

Emboldened by this success, Ffirth moved on to a human subject, himself:

> On the 4th of October, 1802, I made an incision in my left arm, mid way between the elbow and wrist, so as to draw a few drops of blood; into the incision I introduced some fresh black vomit; a slight degree of inflammation ensued, which entirely subsided in three days, and the wound healed up very readily.

Ffirth's experiments grew progressively bolder. He made deeper incisions in his arms, into which he poured black vomit. He dribbled the stuff into his eyes. He cooked some on a skillet and forced himself to inhale the fumes. He filled a room with heated regurgitation vapors—a vomit sauna— and remained there for two hours, breathing in the air. He experienced a "great pain in my head, some nausea, and per-spired very freely," but otherwise was okay.

He then began ingesting the vomit. He fashioned some of the black matter into pills and swallowed them. Next, he mixed half an ounce of fresh vomit with water and drank it. "The taste was very slightly acid," he wrote. "It is probable that if I had not, previous to the two last experiments, accustomed myself to tasting and smelling it, that emesis would have been the consequence." Finally, he gathered his courage and quaffed pure, undiluted black vomit. Having apparently acquired a taste for the stuff, he even included in his treatise a recipe for black-vomit liqueur:

> If black vomit be strained through a rag, and the fluid thus obtained be put in a bottle or vial, leaving about one-third part of it empty, this being corked and sealed, if set by for one or two years, will assume a pale red colour, and taste as though it contained a portion of alkahol.

Despite his Herculean efforts to infect himself, Ffirth had still not contracted yellow fever. He momentarily con-sidered declaring his point proven, but more yellow-fever-tainted fluids remained to be tested: blood, saliva, perspira-tion, and urine. So he soldiered on, liberally rubbing all of these substances into incisions in his arms. The urine pro-duced the greatest reaction, causing "some degree of inflam-mation." But even this soon subsided. And he was still dis-ease free.

Ffirth now felt justified in declaring his hypothesis proven. Yellow fever *had to be* noncontagious. Unfortunately, he was wrong. We now know that yellow fever is caused by a tiny RNA virus spread by mosquitoes. This explains why Ffirth observed seasonal variations in the spread of the disease. The epidemic retreated in winter as the mosquito population lessened.

How Ffirth failed to contract the disease is a bit of a mystery, considering he was rubbing infected blood into wounds on his arms. Christian Sandrock, a professor at UC Davis and an expert on infectious diseases, speculates that he simply got lucky. Yellow fever, much like other mosquito-borne diseases such as West Nile virus, requires direct transmission into the bloodstream to cause infection. So Ffirth happened to pick the right virus to smear all over himself. Had he done the same thing with smallpox, he would have been a goner.

Although Ffirth made a bad guess about the cause of the disease, his experiments weren't entirely in vain. He submitted his research to the University of Pennsylvania to satisfy the requirements for the degree of Doctor of Medicine, which was subsequently granted to him. Modern graduate students who complain about the excessive demands of their thesis committees might want to keep his example in mind. They don't realize how easy they have it.

Ffirth, S. (1804). *A treatise on malignant fever; with an attempt to prove its non-contagious nature.* Philadelphia: Graves.

Magic Feces

Imagine a dog turd. Some unknown pooch deposited it on a lawn weeks ago. Since then it's been baking in the sun until it's formed hard, crusty ridges. Would you want to pick this up and eat it? Of course not.

Now imagine it's 1986 and you're an undergraduate at the University of Pennsylvania. You volunteered to participate in a food-preferences study, and you find yourself sitting in a small, square laboratory room. You've just been given a piece of fudge to eat, and it was very good. The researcher now presents you with two more pieces of fudge, but there's obviously a trick. The fudge morsel on the left is molded into the form of a disk, but the one on the right is in the shape of a "surprisingly realistic piece of dog feces."

"Please indicate which piece you would prefer," the researcher asks in a serious tone.

This question was posed to people as part of a study designed by Paul Rozin, an expert in the psychology of disgust. The responses his team got were no surprise. Participants overwhelmingly preferred the disk-shaped fudge, rating it almost fifty points higher on a two-hundred-point preference scale.

The researchers exposed subjects to a variety of gross-out choices. In each situation the options were equally hygienic—there was never any risk of bacterial infection—but one was always distinctly more stomach turning than the other.

They offered volunteers a choice between a glass of apple juice into which a candleholder had been dipped, or one in which a dried sterilized cockroach had been dunked. "Which would you prefer?" the researcher asked as he dropped the cockroach into the glass and stirred it around. The roach juice scored one hundred points lower on the preference scale.

Would volunteers prefer to hold a clean rubber sink stopper between their teeth, or a piece of rubber imitation vomit? The sink stopper won out.

Would they be willing to eat a bowl of fresh soup stirred by an unused fly swatter, or poured into a brand-new bedpan? "No, thank you," participants responded in both cases.

The researchers offered the results of their study as

evidence of the "laws of sympathetic magic" at work in American culture. These laws were named and first described by the nineteenth-century anthropologist Sir James Frazer in his classic work *The Golden Bough*, an encyclopedic survey of the belief systems of "primitive" cultures around the world.

Frazer noticed two forms of belief showing up again and again. First, there was the law of contagion: "Once in contact, always in contact." If an offensive object touches a neutral object, such as a cockroach touching some juice, the neutral object becomes tainted by the contact. Second, there was the law of similarity: "The image equals the object." Two things that look alike are thought to share the same properties. A voodoo doll that looks like a person becomes equivalent to that person. Or fudge that looks like feces is as disgusting as feces.

In the modern world we like to think of ourselves as being quite rational. We understand the principles of good hygiene, and how diseases are transmitted. We like to imagine we're not ruled by simple superstitions. And yet, of course, we are. As the researchers put it, "We have called attention to some patterns of thinking that have generally been considered to be restricted to preliterate or Third World cultures." The curious thing is that most of us will readily acknowledge the illogic of these beliefs. We know the laws of sympathetic magic aren't real. But we're still not about to eat that doggy-doo fudge.

Rozin, P., L. Millman, & C. Nemeroff (1986). "Operation of the Laws of Sympathetic Magic in Disgust and Other Domains." *Journal of Personality and Social Psychology* 50 (4): 703–12.

Space Invaders in the Loo

A man sits in a stall at a public lavatory on a college campus. He has been there for well over an hour. At his feet is a stack of books, and hidden among these books is a small periscope he is using to peer beneath the door and watch a man standing at a urinal, going to the bathroom. Through his periscope the stall-sitter can see the stream of urine trickling down the porcelain into the drain. The urine stops, and he immediately presses a button on a stopwatch he holds in his hand.

The man in the stall was not a Peeping Tom. He was actually a reputable researcher conducting a scientific experiment. At least that was his story, and he was sticking to it.

The location of this peculiar scene was "a men's lavatory at a midwestern U.S. university." The date was sometime in the mid-1970s. Let's momentarily follow a hypothetical person, probably a student at the university, who wandered into the lavatory and unwittingly became a participant in the stall-sitter's experiment. We'll call him Joe.

Feeling some pressure in his bladder, Joe ducks into the toilets. He sees two stalls and three urinals that extend up from the floor. One of the stalls is occupied by someone who has piled his books at his feet—the covert watcher. But Joe ignores the stalls and heads straight for the urinals. A man is standing at the middle urinal, and a sign hanging on the rightmost urinal reads, "Don't use, washing urinal." So Joe walks up to the leftmost one.

He stands next to the other man—sixteen inches away from him, to be exact—and unzips his fly. Unbeknownst to him, at that moment the man sitting in the stall presses a button on his stopwatch.

Joe can sense the presence of the guy standing next to him. He is, perhaps, a little too close for comfort. But Joe really has to go, so he concentrates, tuning out the thought of the stranger's proximity. At last his bladder muscles relax and the urine starts to flow.

Peering through his periscope, the researcher in the stall sees the stream of urine begin to trickle downward. He presses another button on the stopwatch.

Eighteen seconds later Joe is finished. (The watcher in the stall presses another button.) Joe zips up his fly and walks to the sink to wash his hands. He notices the man at the middle urinal still hasn't finished. "Poor guy," he thinks, "must be having trouble going." Then he dries his hands and leaves, blissfully unaware that he just took part in an experiment.

This scene, with a few variations, played out sixty times at the midwestern university, as different unwitting subjects wandered in and out of the restroom.

The purpose of all this clandestine loo activity was to determine whether "decreases in interpersonal distance lead to arousal as evidenced by increases in micturition delay and decreases in micturition persistence." Translated into plain English: Do guys have more difficulty peeing when they feel crowded? On a nonscientific level, most men would say yes, obviously this is true. Pop culture offers a variety of terms to describe the phenomenon—"choking at the bowl" or "stage fright." But the researchers wanted empirical data, not hearsay. If they could establish a connection between crowding and "onset of micturition," this would demonstrate, they believed, that the body reacts with signs of stress when strangers get too close or invade our personal space.

The experimenters manipulated conditions in the lavatory to force subjects to urinate while standing at varying distances from a stranger. In the "close distance condition," one of the experimenters pretended to use the middle urinal while the DON'T USE sign hung on the rightmost one, forcing subjects to use the one immediately to the experimenter's left. In the "moderate distance condition," he stood at the rightmost urinal with the "don't use" urinal between him and the subjects. In the control condition, signs hung on two of the three urinals, allowing people to pee in relative privacy. Meanwhile the researcher in the stall timed how long subjects took to start urinating, and the duration of their urination.

The results revealed that "closer distances led to increases in micturition delay and decrease in micturition persistence." When guys had to stand shoulder to shoulder with a stranger, they waited, on average, 8.4 seconds before their urine started to flow. They were done 17.4 seconds later. At a moderate distance they fared slightly better—a 6.2-second wait and a 23.4-second duration. But in the control condition, the wait lasted only 4.9 seconds and subjects enjoyed a leisurely 24.8-second pee. This means that for a guy with a full bladder, a crowded bathroom can mean 3.5 extra seconds before the arrival of relief. And sometimes those 3.5 seconds seem like a *loooong* time.

When published in the prestigious *Journal of Personality and Social Psychology,* this study was not universally well received. Gerald Koocher of the Harvard Medical School wrote in, blasting it as "laughable and trivial." He also expressed fear that its publication would encourage a "veritable flood of bathroom research." We can now say that this fear was unjustified. It has, in fact, been more of a slow but steady trickle of research, inhibited perhaps by all those strangers standing close by, waiting and watching.

_____

Middlemist, R. D., E. S. Knowles, & C. F. Matter (1976). "Personal Space Invasions in the Lavatory: Suggestive Evidence for Arousal." *Journal of Personality and Social Psychology* 33: 541–46.

Communal Peeing in Ants

The rain begins to fall in the Malaysian rain forest. Water droplets tumble through the thick, humid air, onto the lush vegetation below. Leaves shake as the downpour hits. Giant bamboo stalks sway with the force of the storm. Inside the internodes of the giant bamboo, the rain is flooding a nest that ants from the species Cataulacus muticus *have spent years building. The nest must be saved. The worker ants mobilize, squeezing their bodies into the entrances, blocking the water with their heads. But it's not enough. Water still leaks in. Luckily, the others know what to do. Hours later the rain has passed, and the nest is bone dry.*

After studying ants in the Malaysian rain forest for years, Joachim Moog and Ulrich Maschwitz thought they knew a lot about ant behavior. But *Cataulacus muticus* was a puzzle to them. How did this ant get the water out of its nest?

Many ant species transport water by holding it in their mouths, then spitting it out. Other ants carry droplets on their backs. But Moog and Maschwitz didn't observe this behavior. Something else seemed to be going on. Something far weirder.

To solve the mystery, the researchers brought a *Cataulacus muticus* ant colony back to their lab at Frankfurt University and subjected it to experimental flooding. They injected two milliliters of yellow-dyed water directly into the nest. Like students at a party, the ants immediately began to drink as much of the yellow liquid as they could. Twenty minutes later they exited the nest en masse. Moog and Maschwitz then observed:

> They moved sideward for several centimeters and raised their gasters steeply. Immediately a clear droplet appeared at the gaster tip which rapidly grew and fell down within a few seconds.

The ants were peeing the water away. To make sure they were seeing this correctly, the researchers repeated the experiment, with the same results. They calculated that 3,030 pee runs were required to dry out the nest.

Such "cooperative peeing behavior" had never before been observed in ants. Moog and Maschwitz tested other species, but found none that employed the same strategy. *Cataulacus muticus* appears to be the only ant species that uses its bladder for flood control.

The communal peeing of *Cataulacus muticus* truly is a wonder of evolution. That an ant species developed such an ingenious nest-saving technique boggles the mind. Unfortunately, the same cannot be said of the communal peeing of students.

———

Maschwitz, U., & J. Moog (2000). "Communal Peeing: A New Mode of Flood Control in Ants." *Naturwissenschaften* 87: 563–65.

The Sweet Smell of Diapers

Feces are disgusting. People don't like coming into contact with them, and will in fact go to great lengths to avoid them. This disgust protects us from bacterial infection. But why, psychologists Trevor Case, Betty Repacholi, and Richard Stevenson wondered, doesn't disgust prove to be more of an obstacle to child care? Babies may be cute and lovable, but they're also prodigious poop machines. Why don't mothers recoil in loathing at the thought of getting up close and personal with another person's excrement? Perhaps, the researchers theorized, some kind of "source effect" modifies the disgust reaction. Perhaps feces (and other unpleasant substances) inspire less disgust if they come from sources familiar to us, such as our own child.

To test this theory, the researchers recruited thirteen mothers to participate in a "Baby Smell Study." This description was slightly euphemistic. The task asked of each subject was actually to sniff dirty diapers, both of her own baby and a stranger's baby, and rate which she found less offensive.

Before the experiment began, the mothers submitted diapers freshly soiled by their babies. The experimenters had stockpiled dirty diapers from a control baby. These were stored in a refrigerator to keep them fresh, but were taken out two hours before the test and allowed to warm to room temperature. By the time the sniffing began, they were good and rank.

To ensure the mothers couldn't identify the diapers by sight, the experimenters placed each diaper in a covered plastic bucket. The aroma wafted up through a hole in the lid. Mothers put their noses right up to the hole and took a good whiff.

The mothers participated in a total of three trials. In the first trial the diaper-bearing buckets were unmarked. In the second and third trials they bore labels identifying the diapers either as those of the mother's child or of "someone else's baby." However, in one of the trials these labels were incorrect.

After smelling the stinky diapers, the mothers rated each odor on how disgusting they thought it was. The results were unequivocal. Mothers preferred the smell of their own baby's poop. In all three of the conditions—unlabeled, correctly labeled, and incorrectly labeled—the mothers rated the odor of their own baby's dirty diapers as less offensive than that of the other diapers. Surprisingly, this preference was most pronounced in the blind trial, where the mothers had no clue which diaper was which.

The preference was so clear-cut the experimenters briefly worried that the control baby's diapers might have been unusually stinky. But the experimenter who handled and prepared the soiled diapers assured them this was not the case. He insisted the odor of all the diapers was "similarly intense and overpoweringly unpleasant."

The experimenters offered two reasons why a mother prefers the smell of her own baby's poop. Either they become used to the smell through repeated contact, or they are able to detect "some quality that signals relatedness." Case, Repacholi, and Stevenson left it to future researchers to provide further clarification.

This experiment ultimately offers a reassuring message:

No matter how stinky, ugly, or disgusting we are, one person will always think we're great—our mother.

Case, T. I., B. M. Repacholi, & R. J. Stevenson (2006). "My Baby Doesn't Smell as Bad as Yours: The Plasticity of Disgust." *Evolution and Human Behavior* 27: 357–65.

Fart-ology

Sometime in the early stone age, the first joke is about to be told.

A small group of cavemen creeps through a forest, clubs in hand. Danger lurks everywhere. They must constantly be on guard. Suddenly the caveman in the lead stops and signals his companions to be silent. They all freeze in place, straining to hear any noises made by a predator. The leader looks slowly back and forth. He signals the others to listen. And then he lets one rip. Caveman guffawing ensues.

As the story of the flatulent caveman illustrates, farts have always been the butt of jokes (so to speak). The merriment that surrounds them has tended to inhibit serious research. However, there are people who make a living studying farts. In vain they point out that excessive flatulence causes extreme discomfort and distress. Someone has to study the problem, no matter how amusing it seems to everyone else. To paraphrase a joke of more recent vintage, for most a fart is just a fart, but for fart doctors it's their bread and butter.

As a sign of the slow advance of fart studies, not until 1991 did researchers precisely determine the normal amount of flatulence produced by healthy subjects in one day. A study conducted at the Centre for Human Nutrition at the University of Sheffield recruited ten volunteers (five men and five women) each willing to live with a "flexible gas impermeable rubber tube" inserted forty millimeters into his or her anus for twenty-four hours. The tube, held in place by surgical tape, led to a plastic bag, from which no gas could escape. The researchers made sure of this:

> The competence of this gas collection system was validated in two volunteers who submerged the lower parts of their bodies in warm water for an hour during which time there were no detectable leaks (bubbling) and gas was collected in the bags.

The subjects ate a normal diet supplemented by two hundred grams of baked beans, to ensure flatus production. Whenever the need to defecate arose, subjects closed off their bag, removed the tube, did their business as quickly as possible, and reinserted the tube. After a day of collection, the gas volume was measured. The median volume came out to 705 milliliters. An average of eight episodes of flatulence were reported. This translated to a median volume of 90 milliliters per episode. The women and men expelled equivalent amounts, proving the equality of the sexes, at least in this matter.

Even more shocking, it was not until 1998 that science identified the exact gases responsible for flatus odor. Dr. Michael Levitt of the Minneapolis Veterans Affairs Center used the same rectal-tube-and-bag system as the 1991 study to collect flatus from sixteen healthy subjects who ate pinto beans the night before to enhance production. Samples were then drawn from the bags via syringe and given to two judges to rate for intensity:

In an odour-free environment, the judges held the syringe 3 cm from their noses and slowly ejected the gas, taking several sniffs. Odour was rated on a linear scale from 0 ("no odour") to 8 ("very offensive").

And you think your job sucks.

Odor intensity correlated with high levels of sulfur gases: hydrogen sulfide, methanethiol, and dimethyl sulfide. These gases were isolated and presented individually to the judges, who described them, respectively, as "rotten eggs," "decomposing vegetables," and "sweet." Therefore Levitt could positively identify hydrogen sulfide as the gas that causes real stinkers. The sickly sweet kind of wiffies, by contrast, are due to an excess of dimethyl sulfide.

Intriguingly, Levitt's study did find a difference between men and women. The women's farts "had a significantly higher concentration of hydrogen sulfide ($p<0.01$) and a greater odour intensity ($p<0.02$) than did that of men." But the men held their own by producing a greater volume of gas overall. So, for now, we can still call this battle of the sexes a draw.

Suarez, F. L., J. Springfield, & M. D. Levitt (1998). "Identification of gases responsible for the odour of human flatus and evaluation of a device purported to reduce this odour." *Gut* 43: 100–4.

CHAPTER NINE

Making Mr Hyde

Human nature has two sides—good and evil. What causes one side to grow stronger than the other? For Robert Louis Stevenson's character Dr. Henry Jekyll, it was a salt containing an "unknown impurity." When Jekyll mixed this salt into a solution and drank it, he transformed into the murderous Mr. Hyde. Many real-life researchers have shared Jekyll's—and Stevenson's—fascination with humankind's wicked ways. They have studied what causes people to become rude, antisocial, overly aggressive, and cruel. Unnervingly, the answers they come up with rarely involve anything as elaborate as crystalline salts. Scientists have discovered that to bring out the worst in a person, it usually suffices to place him or her in the right situation. As Philip Zimbardo, whom we shall meet later in this chapter, once observed, "Any deed that any human being has ever done, however horrible, is possible for any of us to do—under the right or wrong situational pressures." Of course, the same is true in reverse. Given the right situation, any person can be turned into a saint. And many researchers do study the causes of altruistic behavior. But let's be honest—the villains are always more interesting.

Shocking Obedience

A nervous-looking man in a tight-fitting white T-shirt leans forward and speaks into the microphone. "Learner, what is your answer?"

There is no reply. After a few seconds, the man repeats the question more forcefully, "Learner, what is your answer?"

Suddenly a voice shouts through the wall, "I refuse to answer. Let me out of here."

"You've got to answer, otherwise you get a shock."

"I won't answer. You can't hold me here. Get me out. Get—me—out—of—here."

The man turns in his chair and gazes imploringly at the lab-coat-wearing researcher seated behind him. "I don't think he's going to answer."

The researcher calmly replies, "If the learner doesn't answer in a reasonable time, consider the answer wrong."

"But he's yelling in there. He wants out."

"Please continue."

"Maybe we should check in on him. He said that he had a weak heart."

"The experiment requires that you continue."

The man sighs and turns back around. He stares at the instrument panel in front of him. The panel displays a row of thirty switches. Each switch is marked with a voltage level, from 15 volts on the far left, progressing upward in increments of 15 to 450 volts on the far right. Beneath the switch marked 315 volts is a warning: EXTREME INTENSITY SHOCK. The man carefully places his finger on this switch. Then he removes it. Once again he turns to face the researcher.

"I don't want to be responsible for killing a man."

"The responsibility is mine. Please go on."

The man shakes his head as though unsure. A hollow, fearful look flickers through his eyes. He shrugs his shoulders, turns back around, and mutters, "Well, that's that."

He leans forward and speaks into the microphone, "Learner, your answer is wrong." Then he presses the switch. A bloodcurdling scream shakes the walls.

Would you torture or kill an innocent victim on the command of a stranger? When asked this question, almost everyone says no. But almost everyone is wrong. Stanley Milgram's obedience experiment, conducted at Yale University in the early 1960s, demonstrated that the average person is capable of doing horrendous things, especially when told to do so by someone wearing a white lab coat.

Milgram dreamed up his experiment while thinking about the Holocaust. Why was it, he wondered, that German citizens obeyed orders to send millions of Jews to death camps? Was there some quirk in the German character that made these citizens peculiarly obedient to authority? Or is such obedience a common feature of human psychology? If ordered, would Americans have done the same thing? To find out, Milgram decided to place randomly chosen subjects into a situation in which an authority figure would ask them to commit increasingly repellent acts of cruelty. The researcher would not coerce them. Subjects could stand up and leave at any time without consequence. Only a verbal command would be given: "Please go on . . . The experiment requires that you continue . . . You have no other choice, you must go on." How would people respond to this request?

Milgram's subjects were utterly ordinary people—postal

clerks, teachers, salesmen, factory workers. He recruited them by placing an ad in a newspaper, offering four dollars to anyone willing to participate in an hour-long "scientific study of memory and learning."

When a subject showed up at the Yale Interaction Laboratory, where the experiment was conducted, he was led through a series of elaborately staged events. First, a young, serious-looking researcher met him and introduced him to a man described as a second volunteer—a pleasant-looking, round-faced accountant in his late forties. Both the researcher and the second volunteer were actors who had carefully rehearsed the parts they would play during the next hour. Milgram was hidden behind a one-way window, observing everything that happened.

The young researcher provided the subject with a false explanation of the experiment. He said it was designed to examine the effect of punishment on learning. One volunteer would serve as a "learner." He would attempt to memorize a series of word pairs. The other would be a "teacher." He would read the word pairs to the learner. The researcher stressed the next point—the teacher would operate a shock generator. Each time the learner gave a wrong answer, the teacher would administer punishment by flipping a switch on this machine and giving the learner a shock. The shocks would increase in intensity each time an incorrect answer was given.

The two volunteers drew straws to determine who would be the learner and who the teacher. The fake volunteer always got to be the learner. The researcher then made a show of strapping the learner into an electric-chair apparatus—applying electrode gel to his wrists and tightening the restraints to prevent movement. Looking nervous, the learner asked whether the shocks could aggravate a heart condition he had. The researcher dismissed this concern: "Although the shocks can be extremely painful, they cause no permanent tissue damage."

Next, the researcher led the subject into an adjacent room where the voltage panel was housed and showed him how to operate the machine. The teacher settled himself in front of the panel. The researcher sat behind him, across the room, and the experiment began.

It always started calmly. The teacher read out a series of word pairs: *blue/box, nice/day, wild/duck*. Then he read the first word of one of the pairs along with four other terms. *Blue: sky, ink, box, lamp*. He waited for the learner to identify the corresponding term.

The learner aced the first few pairs. The subjects must have imagined there would be no need to explore the horrors waiting at the far right of the panel. But the word pairs became more challenging, and the learner began making mistakes. One error followed another. "Incorrect," the teacher would say, and flip a switch on the voltage panel. Next time the shock was slightly stronger. The time after that it was stronger still.

When the teacher pressed the 75-volt switch, the learner let out a distinct "Ugh" that could be heard through the wall. At 120 volts the learner's reaction became more animated. "Hey, this really hurts," he shouted. By 150 volts the learner was screaming to have the experiment stopped and to be let out. The cries of the learner came from a tape recorder. No one was actually being shocked. But the teachers didn't know that. For them, the screams were terrifyingly real.

Many of the teachers began to sweat and tremble. They bit their lips and dug their fingernails into their palms. Some of them laughed hysterically. All of them looked to the experimenter for guidance. *What should I do now?* The researcher offered calm reassurances and urged them to proceed. "Please go on," he would say. "The experiment requires that you continue."

This was the moment of truth. How far up the panel would the teacher progress? Would he go to 200 volts? 300? 400? When would he push back his chair and say, "No more"? Or would he never do this? Would he press the switches all the way up to 450 volts?

Before he conducted the experiment, Milgram anticipated that virtually no one would go all the way to the end of the panel. Psychiatrists he polled agreed with this prediction. They forecasted that only one subject in a thousand would administer the highest shock. But the actual behavior of the subjects shattered these expectations. *Almost two-thirds of the teachers never disobeyed the experimenter.* They agonized and sweated and shook, but they kept pressing the switch. They pressed the switch after the learner started screaming, after he yelled out that his heart was weak, and after he screamed in agony to be let out. They kept pressing the switch after the learner received 330 volts and fell into an eerie silence, apparently unconscious or dead. They pressed the switch all the way up to 450 volts, and then they kept pressing it until, finally, the researcher told them to stop. These were not serial killers or sadists. These were just average Americans, who were apparently willing to kill an innocent person because a man in a white lab coat told them to. Years later, during a CBS *60 Minutes* interview, Milgram glumly concluded:

> I would say, on the basis of having observed a thousand people in the experiment and having my own intuition shaped and informed by these experiments, that if a system of death camps were set up in the United States of the sort we had seen in Nazi Germany, one would be able

to find sufficient personnel for those camps in any medium-sized American town.

Milgram tried numerous variations of the experiment, searching for the limits of obedience. He discovered that the proximity of the victim had a powerful effect on compliance. If subjects could neither see nor hear feedback from the victim, obedience was almost total. If they could hear only a thumping on the walls, compliance was 65 percent. But if the two people were in the same room, and the subject had to physically press the victim's hand onto a metal plate to give him a shock, compliance dropped to 30 percent. Of course, 30 percent is still dismayingly high. Other variables, such as gender, had little effect on the results. Women proved just as willing as men to shock the victim.

Milgram's obedience study offers a depressing view of human nature. The average person seems all too willing to follow orders, no matter how cruel or unjust. But humanity's stock sinks even lower when you consider a similar experiment conducted in Chicago during the same period. The Chicago researchers locked rhesus monkeys in cages. To obtain food, the monkeys had to pull on a chain. But there was a catch. Pulling the chain also caused a monkey in a neighboring cage to receive a high-frequency shock. After witnessing the agony of their neighbors, the majority of the monkeys refused to pull the chain again. They starved, some for as long as twelve days, instead of inflicting pain on another. The monkeys, in other words, did something most humans could not: They said no. Apparently we still have much to learn from our primate cousins.

Milgram, S. (1974). *Obedience to Authority: An Experimental View.* New York: Harper & Row.

Shock the Puppy

When Stanley Milgram published the results of his obedience experiments in 1963, they sent (figurative) shock waves through the scientific community. Other researchers found what he was reporting hard to believe. Could subjects really be so easily manipulated? They were sure Milgram must have made a mistake. Researchers conducted numerous follow-ups to his experiments, searching for ways to bring his results back in line with expectations. One experiment, carried out by Charles Sheridan and Richard King in 1972, easily stands out from this crowd.

Sheridan and King theorized that Milgram's subjects suspected the victim was fake. This would explain their remarkable obedience. They were just playing along with the game. To test this possibility, Sheridan and King decided to repeat Milgram's experiment *using an actual victim who would really get shocked*. Obviously they couldn't use a human for this purpose. So they used the next best thing—a cute, fluffy puppy.

The experimenters placed the puppy inside a box that had a shock-grid floor. The interior of the box contained a signal light. Subjects—all volunteers from an undergraduate psychology course—were told the puppy was being trained to distinguish between a flickering and a steady light. The dog had to stand either to the right or the left depending on the cue from the light. If the animal failed to stand in the correct place, the subjects had to press a switch to shock it. As in the Milgram experiment, the shock level increased fifteen volts for every wrong answer.

The human subjects could not see the light from where they stood. They could only see the position of the puppy. They judged its responses based on a chart they were given.

Sheridan and King stressed the importance of this research, claiming they were attempting to measure "critical fusion frequency (CFF) in puppies," but they also assured the volunteers that they would receive their compensation, which was course credit, simply for having shown up.

The experiment began, and the puppy immediately got a lot of wrong answers. In fact, there was no right answer for the puppy to get. There was no correlation between the signal light and the answer sheet that had been provided to the students. From the puppy's point of view, it was getting shocked randomly.

As the voltage increased, the puppy first barked, then jumped up and down, and finally started howling with pain. The volunteers were horrified. They paced back and forth, hyperventilated, and gestured with their hands to show the puppy where to stand. Many openly wept. Yet the majority of them, twenty out of twenty-six, kept pushing the shock button right up to the maximum voltage. This finding validated Milgram's results.

In their write-up of the study, the experimenters noted that the shocks were amperage-limited and did not cause the puppy any permanent physical harm. However, they made no mention of psychological harm. If the poor creature later shook with terror whenever it came to a traffic light while out on its walkies, you could understand why.

Sheridan, C. L., & R. G. King (1972). "Obedience to Authority with an Authentic Victim." *Proceedings of the Annual Convention of the American Psychological Association* 80: 165–66.

Requiem for a Rat

The young woman holds the white rat in her hand. It struggles to get free, so she grips it tighter. "Do you seriously want me to do this?" she asks. The researcher standing in front of her nods. "But why?" As she says this, the woman suddenly laughs. It's a nervous, awkward laugh, as though she can't believe the situation she's found herself in. "It is important for the experiment that you proceed," the researcher says. The woman's laughter turns into tears that roll down her cheeks. "Please don't make me do this," she begs. "Please . . ." "The experiment requires that you do it," the researcher states firmly.

Decades before Stanley Milgram shocked the world by demonstrating how readily people will obey a repellent order, a young graduate student at the University of Minnesota witnessed a similar phenomenon in his lab.

It was 1924, and the student was Carney Landis. As part of his doctoral research, he was studying facial expressions. He wondered whether every emotion produces a characteristic expression. Is there one expression used by everyone to show fear? Another to show disgust? Another for arousal? And so on.

To find out, he brought subjects one at a time into his lab. He drew lines on their faces with burnt cork, to better observe which facial muscles they were using. The lines made them look a bit like painted tribal warriors. Then he contrived ways to make them experience emotions. As they expressed each emotion, he took their photos.

The situations in which Landis placed his subjects began

with the mundane. They listened to some jazz music. They read the Bible. They told a lie. They smelled ammonia.

Gradually, the situations became more unusual. BANG! A firecracker went off behind a curtain, and the camera snapped as their faces registered shock. Landis brought out pictures of skin-disease patients, pornographic scenes, and artistic nudes. The camera clicked away as the subjects browsed the images.

Next came the mystery bucket. "Reach into it," Landis told them, "and tell me what you feel." Carefully they placed their hands inside. *Ewww.* Their faces wrinkled with displeasure as they touched three slimy frogs sitting in a puddle of water. "Yes, but you have not felt everything yet," Landis said. "Feel around again." They did so and—ZAP!—they received a powerful electric shock from wires attached to the bucket.

But all this was a mere prelude to what came next—the experimental coup de grâce. Landis carried out a live white rat on a tray. "Hold this rat with your left hand," he told them, "and then cut off its head with the knife."

The subjects stared at him in disbelief. They hadn't been expecting this. They questioned whether he was serious. When he assured them he was, they hesitantly picked the knife up and put it back down again. Many of the men swore. Some of the women started to cry. They pleaded with him to stop the experiment. Nevertheless, Landis urged them on. Hovering over the rat with their painted faces, knife in hand, they now looked even more like members of some strange tribe preparing to offer a sacrifice to the Great God of the Experiment.

It took a lot of coaxing, but eventually 75 percent of his subjects—fifteen out of twenty—complied. They decapitated the rats while the animals were still alive and squirming in their hands. This percentage was similar to the obedience levels Milgram would later find in his electric-shock experiments at Yale.

In general, the procedure went badly. Landis noted, "The effort and attempt to hurry usually resulted in a rather awkward and prolonged job of decapitation." Nor did the rats get a reprieve if the subjects refused to obey. In the five cases of noncompliance, Landis simply picked up the knife and did the job himself. He was determined not to let those rats live.

Most of Landis's subjects were fellow graduate students at the University of Minnesota, but Landis also tested a thirteen-year-old boy suffering from high blood pressure. Doctors had referred the boy to the department of psychology because they suspected his symptoms were caused by emotional instability. You have to wonder how much being forced to decapitate a rat added to his issues.

Landis stumbled upon the phenomenon of experimental obedience almost forty years before Milgram, but Landis never realized the significance of what he had found. It never occurred to him that the willingness of his subjects to obey bizarre commands was far more interesting than their facial expressions as they did so. As it turned out, their expressions varied so widely he failed to find any one look that typified a situation. For instance, expressions shown while decapitating a rat included pained smiling, crying, and what Landis called "fascinated attention," produced by "a slight contraction of the risorius, medium contraction of the zygomatics and lowering of the upper eyelids." Landis died in 1962, just as Milgram was conducting his more famous obedience studies.

It is often this way with experiments. A scientist sets out to prove one thing, but stumbles upon something completely different, something far more intriguing. For this reason, good researchers know they should always pay close attention to strange events that occur during their experiments. A great discovery might be lurking right beneath their eyes—or beneath the blade of their knife.

___

Landis, C. (1924). "Studies of Emotional Reactions, II., General Behavior and Facial Expression." *Journal of Comparative Psychology* 4 (5): 447–509.

What a Difference a Bag Makes

Oregon State University, 1967. It is a cold winter day. A car pulls up to the curb. The passenger door opens, and a man enclosed in a black cotton bag stumbles out. Only his feet protrude from beneath the fabric. As he sways back and forth trying to gain his balance, the car pulls away with a screech of the tires. Having steadied himself, the man in the bag proceeds forward with an air of determination. He walks up the stairs of Shepard Hall, through the doors, and into the classroom of Charles Goetzinger. The students in the room turn to stare as he enters. Dr. Goetzinger looks up from his notes. "Ah, welcome back, Bag. Good to see you again."

Most of the students who attended Charles Goetzinger's class, Speech 113: Basic Persuasion, in the winter quarter of 1967 wore normal clothes—shirts, shoes, slacks, or skirts. But one student opted to show up in a large black bag. He shuffled into class on the first day and took a seat at the back. He didn't say a word.

The Black Bag, as he came to be known, showed up for every class. At first he maintained his silence. When members of the class were each required to give a short speech, he stood before his fellow students for four minutes without saying a word, then returned to his seat. Eventually, as the quarter wore on, he loosened up and let fly with a few cryptic remarks such as, "I'm not Jesus Christ or anything. I'm just one of you in a bag." Reportedly, the Black Bag spoke with a New England accent.

More curious than the man in the bag, though, was the reaction of people to him. Initially the other students tried to ignore him. However, that proved impossible. Although he did nothing to impose himself on others—he barely even spoke—his mere presence dominated the room and seemed to antagonize people. One student threw punches at him. Another tried to pin a KICK ME sign to his back. When the Black Bag responded by sitting next to that student and staring at him, the student poked the Black Bag with his umbrella and screamed, "Get away from me!"

The media picked up on the story of the Black Bag, and reporters soon were descending on the classroom in droves, at one point outnumbering the students. The reaction of the American public resembled that of the students. It was a mixture of curiosity and anger. People wrote letters demanding Goetzinger be fired for allowing such a charade. One alumnus declared that OSU had degenerated to the level of UC Berkeley.

Why did the Black Bag inspire such anger? Professor Goetzinger had one theory: "We always have a frame of reference for events. Then in walks a black bag with a human inside it. Nowhere in our frame of reference has there been such a thing. So we resent."

An experiment conducted a year later on the other side of the country offers another clue.

Philip Zimbardo, a professor at New York University, was

researching the concept of deindividuation. According to this theory, our sense of social responsibility is strongly tied to our feelings of individuality. In situations where we lose our sense of being a unique, identifiable person—for instance, if we blend into a crowd or make ourselves anonymous by putting a bag over our head—we suddenly feel freer to engage in behavior considered antisocial or taboo. Members of the Ku Klux Klan, for instance, both rampage around in mobs and conceal their identities with hoods.

Zimbardo demonstrated the phenomenon by asking two groups of college coeds to electroshock an innocent victim after they had listened to a tape-recorded interview with her. He made the first group feel anonymous by placing large bags (flower-patterned pillowcases) over the participants' heads—making them look, oddly enough, like Klan members. He addressed them by number (never by name), sat them in darkened cubicles, and told them he would not know which one of them was pressing the shock button because all their buttons led to a common terminal (which was a lie). With the second group, he emphasized the subjects' identities. He addressed each one by name and gave them large name tags to wear instead of bags.

The bag-wearing coeds held down the shock button far longer than the non-bag-wearing group. In fact, they often jammed down the button for as long as they could, despite the screams and cries of the victim—who was acting and never got shocked. "These sweet, normally mild-mannered college girls," Zimbardo observed, "shocked another girl on almost every one of the twenty trials on which they had an opportunity to do so, sometimes for as long as they were allowed, and it did not matter whether or not that fellow student was a nice girl who didn't deserve to be hurt." Wearing a bag—even a flower-patterned one—had unleashed their most violent, antisocial impulses.

But while the bag-wearing subjects of Zimbardo's experiment were the ones being aggressive toward another person, at OSU it was the guy in the bag who was suffering abuse. How do we account for this difference?

Zimbardo suggests that the phenomenon of deindividuation works in both ways. Anonymity loosens the restraints on aggressive behavior in situations that permit such behavior, but when victims are anonymous, and therefore dehumanized, it similarly grows easier to commit violence against them. Zimbardo notes numerous reports of children at Disneyland striking hapless costumed characters for no apparent reason.

The idea that we are aggressive toward those with whom we share no comparable identity might also explain what happened as the academic quarter at OSU neared its end. The students warmed up to the Black Bag. They went from bullying him to becoming his strongest supporters, vigorously defending his right to wear a bag. Evidently, in their eyes, he had acquired a recognizable, albeit eccentric, identity. And though he may have been a bag, he was *their* bag. One of them even declared, "If my mother tried to take that bag off him, I'd beat the hell out of her."

When the quarter ended, the Black Bag slipped back into the shadows without revealing his identity. To this day who he was is unknown. So why did he wear the bag? One theory is that it was a stunt inspired by a cryptic remark once made by Anthony Cox, the second husband of Yoko Ono. He had said, "The world would be better off if everyone wore a large, black, cloth bag." Another theory is that Goetzinger put the Black Bag up to it as an informal experiment in persuasion—which was, after all, the theme of the class. One OSU faculty member, convinced this was the case, openly criticized Goetzinger for not including proper sociological controls in his study. However, if this was an experiment in persuasion, it's not clear what the Black Bag was trying to persuade anyone of. That, to paraphrase Winston Churchill, remains, like the Black Bag himself, a riddle inside an enigma wrapped in a mystery covered by a black cotton sack.

Zimbardo, P. G. (1969). "The Human Choice: Individuation, Reason, and Order Versus Deindividuation, Impulse, and Chaos." In *Nebraska Symposium on Motivation* (vol. 17), eds. W. D. Arnold & D. Levine. Lincoln: University of Nebraska Press.

Behind the Wheel

The light turns green. Your foot hovers over the accelerator, but the car in front of you hasn't moved. A few seconds pass and you think, "What is this idiot doing? Why do they let people like this on the road?" A few more seconds go by, and your hand reaches for the horn. "Come on, buddy! I haven't got all day." You start shouting, even though no one can hear you. "Move it!" The anger flows through you, out of your hand, and into the horn. The blaring honk becomes your scream of rage, the sonic weapon with which you assault the driver blocking your path. You allow it to blast on and on. Inside the other car, the driver looks at his watch and chuckles.

If you've ever been stuck behind an unmoving car at a green light, you know the feeling. You can't help but suspect the driver of the stationary car is sitting there on purpose, just to annoy you. As it turns out, you might be right.

Sitting at green lights until people start honking is a favorite experimental technique of anger researchers. When anger experiments are performed in a lab, it's difficult to guarantee subjects will act naturally. They know they're being observed, so they're on their best behavior. But in the naturalistic setting of a traffic intersection, people don't expect a scientist to be watching them. Researchers can study their unguarded reactions, investigating exactly which variables promote an anger response and which inhibit it. This setting also allows researchers a speedy getaway, in case subjects become a little too mad.

Anthony Doob and Alan Gross conducted the first such traffic-intersection experiment in 1968. Drivers in Palo Alto and Menlo Park were their unlucky, unwitting subjects. Doob and Gross wanted to know whether signs of wealth and high status would inhibit displays of aggression. So they drove around, pausing for an obnoxious length of time at green lights, in two separate cars—a black 1966 Chrysler Crown Imperial hardtop, recently washed and polished, and a decrepit gray 1961 Rambler sedan. They recorded which car got honked at more often.

The Rambler won the honk contest, hands down. Eighty-four percent of drivers stuck behind it honked within twelve seconds. This compared to a honk rate of only 50 percent from cars behind the Chrysler. In fact, two impatient drivers never even bothered to honk at the Rambler—they nudged it with their bumpers instead. On the occasions this occurred, the researchers decided waiting around for a honk wasn't wise. They immediately implemented the speedy getaway option.

Since Doob and Gross's study, numerous researchers have performed variations of the honking study. For instance, a 1971 study by Kay Deaux revealed that both genders honk at women drivers more often, apparently because "common stereotypes regarding 'damn women drivers' suggest that honking at a female driver is more acceptable than honking at a male driver." A 1975 University of Utah study discovered that displaying a hostile symbol such as a rifle in a gun rack or a vengeance bumper sticker increases the likelihood people will honk at you.

In one particularly memorable study, conducted in 1976, researcher Robert Baron arranged for a male accomplice to stop at a red light. When a car pulled up behind him, and with the light still red, a female accomplice walked across the street in between the two cars. She did so either dressed conservatively in blue jeans and a blouse, limping on crutches, wearing a clown mask, or attired in a sexy, revealing outfit. After she crossed the street, the light turned green, and the accomplice in the car waited fifteen seconds before moving. Baron found that when the female accomplice had just walked by on crutches, wearing a clown mask, or dressed provocatively, people in the second car honked significantly less often than when she had passed by in conservative clothes or when she didn't cross the street at all. These results, he suggested, indicate that empathy, humor, and mild sexual arousal may inhibit aggression.

While contributing to this field of research may be tempting—especially given the simplicity of the procedural design—this is one experiment that should definitely be left to the professionals. The last thing our roads need are scores of amateur researchers blocking intersections in the name of science. Except, perhaps, for the skimpy bikini test. That one might be worth encouraging.

———

Doob, A. N., & A. E. Gross (1968). "Status of Frustrator as an Inhibitor of Horn-Honking Responses." *Journal of Social Psychology* 76: 213–18.

The Stanford Prison Experiment

"Prisoner 8612, against the wall!" The prisoner ignores the guard. His mind is reeling. It feels like everything is pressing in on him, as though he is going mad.

"Against the wall!" the guard shouts again. "Come on, somebody get him back in line."

Suddenly 8612 wheels to face the guard. "Listen, if I have to be in here, I'm not going to put up with this shit. I mean, really!" The prisoner turns and grips one of his fellow prisoners by the arm. "I couldn't even get out," he hisses. There's a desperate edge to his voice. "They wouldn't let me out. You can't get out of here."

The other prisoners laugh nervously, but you can see it in their eyes—the sudden flash of panic. He couldn't get out? That means that this is real, that this is an actual prison. And they're stuck inside.

It had started out like a game. They had volunteered to spend two weeks in a fake jail in the basement of the Stanford psychology department, dressed up as prisoners and guards. It would be fun. Or, at least, something different to do during the summer. The idea seemed harmless. How bad could it be?

The mock prison was the idea of Philip Zimbardo—the same Zimbardo we just met who dressed New York University coeds in flower-patterned bags. By now it was 1971, and he was teaching at Stanford, but he was still interested in situations that make good people turn bad, and his thoughts had turned to prisons. What makes prisons such violent places? he wondered. Was the character of the prisoners and guards to blame? Or was it the power structure of the prison itself that brought out the worst in people? Was it a case of bad apples, or a bad barrel?

To find out, he decided to create a fake prison. He would recruit twenty-four healthy young men—who were good, honest citizens with no criminal records, and who, based on their personality test scores, fell into the normal range on every trait—and randomly assign half of them to be guards in his mock prison and the other half to be prisoners. He would then step back for two weeks and see what happened.

If bad people are the cause of bad prisons, then his prison filled with good people should experience two uneventful weeks. But if the structure of prison life is responsible for making the inhabitants of prisons go bad, then things would go very differently.

The experiment began on Saturday, August 14, 1971. With sirens wailing, squad cars swept through Palo Alto picking up the prisoners. "Come quietly, son," the officers said, as they led the bewildered young men to the squad cars in full view of concerned neighbors. *Am I being arrested?* the volunteers wondered. They had been told nothing about real police picking them up. Only when they were deposited at the Stanford psychology department did they know, *Yes, this was part of the experiment.* Zimbardo dreamed up the police pickup as a way to thrust the prisoners right into their new roles. The police had gone along as a conciliatory gesture toward Stanford's administration, with whom tensions had recently run high following violent anti–Vietnam War demonstrations.

The prison was a bunch of offices with bars fitted onto the doors. A broom closet served as solitary confinement. The guards, wearing khaki uniforms and reflective sunglasses, met the prisoners. "From now on you are a number," they told them. "You will address each of us as 'Mr. Correctional Officer.' Do you understand?"

The prisoners stripped and were sprayed with deodorant, as if being deloused. Then they were given smocks, stocking caps, and ankle chains to wear. The smocks (worn without underwear) humiliated them by forcing them to walk awkwardly to keep themselves covered. The caps simulated having their heads shaved, and the ankle chains were to remind them of their loss of freedom.

The first day was a lot like being at camp. The guards were unsure how to behave. They had received only minimal instructions—no physical violence, and don't let the prisoners escape. The prisoners seemed more at ease, swapping jokes with one another as they lined up for roll call.

But before long the guards warmed up to their roles. At two a.m. they forced the prisoners out of bed for a count in "the yard" (which was really the hallway outside the offices). "Get out of bed! On the double!" they screamed. "Up against the wall!" Bleary-eyed, the prisoners complied.

The next morning the prisoners retaliated. They staged a rebellion, shoving their beds up against their cell doors and screaming taunts at the guards: "This isn't a prison. This is a fucking simulation!"

Embarrassed by the sudden loss of control, the guards cracked down hard. They blasted the prisoners with fire extinguishers, called in the off-duty guards as reinforcements, and forced their way into the cells. From that moment on, summer camp was over. The guards stripped the prisoners naked, herded them into the yard, and made them do jumping jacks, sit-ups, and push-ups. They threw the ringleader of the rebellion, prisoner 8612, into solitary to let him dwell on his misbehavior.

To prevent future rebellions, the guards sharply curtailed the few liberties the prisoners had. They implemented random strip searches and revoked bathroom privileges, making the prisoners pee in a bucket. Soon the stench of urine crept through the cells. The guards also introduced psychological tactics meant to create divisions among the prisoners. They branded those who resisted their authority as troublemakers and blamed them for making conditions worse for the others; meanwhile, they created a 'privilege cell' for good prisoners.

It all proved too much for prisoner 8612. He began complaining of stomach pains and headaches. He approached the researchers, pleading to be let out. But they were unsympathetic. Zimbardo tried to cut him a deal—no more abuse from the guards in return for sharing information about the other prisoners. Dazed and confused, prisoner 8612 stumbled back to his cell, and it was then he told the other prisoners that he couldn't get out, that this was a real prison.

Later that night, 8612 became unmanageable. He started screaming, "I feel so fucked-up. Jesus Christ, I'm burning up inside. Don't you know? I want out!" Grudgingly, Zimbardo agreed to release him. After a mere thirty-six hours, the experiment had lost its first prisoner.

*

Fast-forward to Friday afternoon, the sixth day of the experiment. Christina Maslach stopped by the prison at the request of Zimbardo. She had recently completed her doctorate at Stanford and was romantically involved with Zimbardo. The two would later marry (and they remain married to this day). Maslach had agreed to interview some of the participants and record their thoughts and feelings at this stage of the experiment.

When she arrived, the prison was quiet. The guards were relaxing and the prisoners were in their cells. She met up with Zimbardo, who excitedly related the events of the previous week. "The psychology of it is fascinating," he enthused.

Since the release of 8612, the guards, intoxicated by the sense of power, had steadily escalated their harassment of the prisoners. While refraining from physical violence, they had freely used verbal abuse, humiliation, sleep deprivation, and withholding of basic necessities such as food and blankets. As though aware that what they were doing was wrong, the guards had attempted to conceal their behavior. They made the prisoners write letters home saying, "No need to visit. It's seventh heaven." They also saved their worst abuses for the night shift, when they thought the researchers weren't observing.

The prisoners, meanwhile, had grown more passive, as though broken by the system. Four more of them had followed 8612's lead and started acting crazy, forcing the researchers to release them. One had even broken out in a full-body stress-related rash.

Zimbardo encouraged Maslach to observe "the count," which was just then beginning in the yard. She watched in horror as the meanest of the guards, a blond eighteen-year-old nicknamed John Wayne, strode back and forth, pounding his billy stick into his hand, screaming abuse at the prisoners. Later she witnessed a bathroom run. Hooded and chained, the prisoners were led in single file to the lavatory like animals.

"What you are doing to these boys is a terrible thing!" Maslach burst out. Enraged, she accused Zimbardo of creating a madhouse. How could he allow this to go on? He defended himself. Couldn't she see what important psychological research was being done here? They went back and forth. As they argued, Zimbardo eventually looked out at the prison and stopped in his tracks. He realized Maslach was right. He *had* created a madhouse. He had allowed himself to get caught up in the same negative psychology of the situation that had, within six days, transformed average college kids into passive prisoners and sadistic guards. It had to end.

Not surprisingly, the prisoners were relieved when he halted the experiment the next morning. The guards, however, were disappointed. Most of them had grown quite fond of their newfound power.

The dramatic nature of the Stanford prison experiment—its demonstration of the speed with which social roles can overwhelm people—has made it one of the most famous psychology experiments of all time. Its main challenger for that title is Milgram's obedience study. Both experiments have had a profound cultural impact. They have inspired books, plays, and movies—most notably *Das Experiment*, a 2001 German film loosely based on the prison experiment, and *Atrocity*, an award-winning 2005 movie that reenacts the obedience experiments. Rock bands have even named themselves after the experiments, such as the Los Angeles band Stanford Prison Experiment, and the French punk group Milgram, which has issued a CD titled *450 Volts*.

However, the experiments are linked at a deeper, more personal level. In an odd coincidence, Milgram and Zimbardo were senior-year classmates at James Monroe High School in the Bronx. Zimbardo credits both of them growing up to become situational psychologists to their both being poor, ambitious kids deeply aware of the external forces shaping their lives. He also remembers that Milgram was considered the smartest kid at the school, whereas he was voted most popular boy in the senior class. If they had been judged by their peers to be *most shocking* and *most likely to end up in jail*, that also would have been correct.

Zimbardo, P. (2007). *The Lucifer Effect: Understanding How Good People Turn Evil*. New York: Random House.

The Unresponsive Bystander

You're seated in a small, nondescript office. The experimenter's assistant hands you a pair of headphones with an attached microphone and indicates you should put them on. You do so, and he gives you a thumbs-up sign. Then he exits the room, leaving you alone. Soon the voice of the lead researcher begins to come through your headphones:

> *Good afternoon, everyone. Thank you for coming. We have invited the six of you here today to share your thoughts about personal problems associated with college life. We are very interested in learning how students adapt to life in the urban environment of New York City. To minimize embarrassment, we have taken a number of precautions. First, as you are already aware, today's discussion will be conducted over an intercom system, instead of face-to-face. Second, I will not*

listen to your initial discussion. This is to ensure that none of you are inhibited by the presence of an outside listener. Because I will not be listening, we have implemented a mechanical system to structure the discussion. Each person will be allowed two minutes to speak. As one person is speaking, all the other microphones will be shut off. Once two minutes are up, the machine will automatically turn on the microphone of the next person. We will go around the group a few times in this way, and then open up the microphones for a free discussion. If everyone is clear about this procedure, then I will turn off my headphones and allow your discussion to begin.

You settle into your chair, thinking about what you want to say. Just then the voice of another participant crackles through the headphones, "Hi everyone. I guess it's my turn first."

He sounds like a pleasant, slightly nervous young man. He describes some of the challenges he's faced adjusting to big-city life. One particular problem, he confesses with evident discomfort, is that under conditions of stress, such as exams, he's prone to seizures. His two minutes end, and the microphone of the next person turns on. The discussion proceeds like this through the entire group, until your turn arrives. You talk about difficulties with your roommates, and about the pressures of balancing work and social life. Then your turn is over, and the microphone of the first guy switches on again.

He begins to talk, but abruptly his voice grows louder and his words start to slur. He seems to be having some kind of problem:

"I-er-um-I think I-I need-er-if-if could-er-er-somebody er-er-er-er-er-er-er give me a little-er-give me a little help here because-er-I-er-I'm-er-erh- h-having a-a-a real problem-er-right now and I-er-if somebody could help me out it would-it would-er-er s-s-sure be-sure be good . . ."

"My God," you think. "He's having a seizure." Thoughts race through your mind. "He needs help. Should I tell the experi-

menters? Has someone else already told them?" As you ponder these questions, you continue to hear the young man's agonized plea in your headphones.

"If somebody would-er-give me a little h-help-uh-er-er-er-er-er c-could somebody-er-er-help-er-uh-uh-uh."

Your heart rate accelerates. Someone has to help him! But another part of your mind restrains you—surely others have already rushed to his aid. You would look silly arriving last on the scene. You would only get in the way.

You hear choking sounds. "I'm gonna die-er-er-I'm . . . gonna die-er-help-er-er-seizure-er." You feel simultaneously panicked and useless. All you can think is, "What should I do? What should I do?"

In 1968 this scene replayed thirteen different times at Columbia University. Each time a single, bewildered listener struggled to decide whether to rush out and help the young man, or wait for someone else to do so.

What these people didn't know was that there was no medical emergency. Nor were they involved in a roundtable discussion about student life. The sound of the man having a seizure, as well as the voices of the other participants, came from a tape recorder. By volunteering to participate in the discussion of problems associated with college life, they had become guinea pigs in John Darley and Bibb Latané's "bystander intervention in emergencies" experiment.

Darley and Latané were investigating the curious phenomenon of unresponsive bystanders—how it is that crowds of people can watch as an emergency unfolds, without a single person stepping forward to help. The direct inspiration for their experiment was the widely reported 1964 murder of Kitty Genovese. As Genovese returned to her apartment at three a.m., she was attacked outside her building by a man with a knife. He stabbed her repeatedly and raped her as she screamed for help. This went on for half an hour. Lights came on in the apartments above. People came to their windows to see what all the noise was about. But no one came to her aid.

Genovese's murder caused a public outcry. The people who did nothing were condemned as apathetic and heartless. But Darley and Latané suspected individual psychology had little to do with the witnesses' lack of response. Instead, they suspected group psychology was to blame. They devised their experiment to prove this.

They led subjects to believe they were either part of a two-person discussion group, or (as depicted in the scene above) a six-person group. In the two-person condition, subjects invariably ran out of the room to alert the experimenters as soon as the young man began to experience the seizure. They knew they had to do something because—so they believed—no one else was able to hear his cries. But subjects behaved very differently in the six-person condition. Now they hesitated before acting. They sat passively, agonizing about what to do, wondering if someone else was going to help. Thirty-eight percent of the subjects never left the room. Just like the witnesses of Kitty Genovese's murder, they became unresponsive bystanders.

The phenomenon the experiment demonstrated is called "diffusion of responsibility." When people in groups witness an emergency, they tend to look around and think, "Someone else will help." No one feels directly responsible. And, consequently, no one does anything.

When Darley and Latané published their results in 1968, the scientific community hailed the experiment as a classic study. In the decades since then, other researchers have extended their work in many ways, staging fake emergencies, including robberies, kidnappings, women being attacked, subway passengers collapsing and drooling blood from their mouths, and men gushing blood from arterial wounds. Any shocking emergency you can think of has probably been simulated somewhere by a social psychologist for the benefit of a horrified crowd.

These studies have taught us a great deal about group psychology and the phenomenon of diffusion of responsibility. But they have had an unintended consequence, because in addition to unresponsive bystanders we now also have to worry about skeptical ones. This was revealed during a 1986 experiment conducted by Robert MacCoun and Norbert Kerr. The two researchers were staging a mock trial. Suddenly a psychology student playing the part of a jury member had a grand mal epileptic seizure—*for real.* But many of the people in the room, familiar with Latané and Darley's study, thought it was just part of the experiment. Even when the paramedics arrived, many were still convinced the student was acting. Luckily the seizure victim did get help and everything turned out fine. But what this incident demonstrates is that if you ever do need help in a crowded place—perhaps you trip and break your leg on a busy street—there's a real danger you may bleed to death before anyone realizes you're not just playing a part in a weird psychology experiment.

Darley, J. M., & B. Latané (1968). "Bystander Intervention in Emergencies: Diffusion of Responsibility." *Journal of Personality and Social Psychology* 8 (4): 377–83.

CHAPTER TEN

The End

We arrive at the end. Not the end of the book—not yet, at least. Rather, the end as the theme of this chapter.

Traditionally, there has been some tension between religion and science regarding matters of jurisdiction when the end arrives, whether we're talking about the end of life or the end of the world. In the early days of experimentation, this dispute often came to a head in cemeteries. A priest would put bodies into the ground, and the grave robbers, paid by men of science, would haul them back out and cart them off to the medical labs. During the nineteenth century this practice became such a problem that many families held vigils at graves to prevent the bodies of their loved ones from being exhumed.

Nowadays it is not just the anatomy of the dead that scientists are curious about. Psychologists study how death motivates our actions in life, and pharmacologists explore how drugs can alter the process of dying. All of these studies are grouped together into the broad interdisciplinary field called thanatology—from the Greek word *Thanatos*, meaning death.

In a way, these studies represent the exact opposite of the Frankenstein experiments with which this book began. Those studies, despite the many corpses they featured, sought to understand the force of life. Here it is death—and the shadow it casts across our lives—that comes under the microscope.

Fear Factor

A prop plane cruises through a clear blue sky. Onboard, passengers lean back in their seats. A few pull out books. Others stare out their windows, anticipating an uneventful flight. But suddenly the plane shudders violently and banks sharply to the left. One of the propellers stalls. The plane begins to spiral slowly downward as the pilot struggles to control it. The passengers hear him in the cockpit, shouting into the radio to the flight controllers on the ground, "We need to make an emergency landing! Repeat. An emergency landing!" People clutch the armrests of their seats so tightly their knuckles turn white. At the back of the plane, a woman starts screaming, "We're going to die! We're all going to die!"

How would you react in such an emergency? Would you remain calm, rationally assessing your best options for survival, or would you be the person screaming hysterically? For the U.S. Army, this question was of more than academic interest. The army needed to make sure soldiers kept their wits about them when bullets started to fly. So, in the early 1960s, it commissioned a team of psychologists—Mitchell Berkun, Hilton Bialek, Richard Kern, and Kan Yagi—to study the phenomenon of "behavioral degradation under psychological stress." The army wanted to know how badly the performance of the average soldier suffered when he thought he was about to die, and whether there were techniques soldiers could learn to help them function more effectively in fear-arousing circumstances.

There's really only one way to find out how people will behave in a life-and-death situation. You have to scare them into believing their lives are in danger—or, as the researchers put it in the dry language of science, you have to effect "the experimental arousal of fear of death." Soldiers going through basic training at the Hunter Liggett Military Reservation in central California became the subjects. Naturally, none of the soldiers were briefed that the terrifying events they would soon experience were part of an experiment. That would have ruined the effect.

The first fear-arousing situation the researchers dreamed up was a flying laboratory of terror. Groups of soldiers were taken aloft in a small propeller plane. When it reached cruising altitude, the plane suddenly lurched and the propeller stalled. Over their headsets the soldiers heard the pilot talking to the tower: "Something's wrong. We have to make an emergency landing." The plane circled to return to the airport, and the soldiers could see ambulances and fire trucks waiting down below. At this sight, a knot of fear must have formed in the men's throats, but then the situation grew even worse. The pilot announced that the landing gear wouldn't come down. He was going to have to attempt to ditch the plane in the ocean.

Having established the fear-arousing situation, the researchers next introduced a task to measure the soldiers' ability to perform under pressure. Somewhat incongruously, the task was to fill out insurance forms. A steward distributed the paperwork, explaining it was a bureaucratic necessity that everyone fill it out—if they were all going to die, the army wanted to make sure it was covered for the loss. The forms were to be placed in a canister and jettisoned before the crash landing. Obediently, the soldiers leaned forward in their seats, pencils in hand, and set to work deciphering the legalese. "These forms are pretty hard to understand," they probably thought to themselves. Perhaps they attributed the difficulty they were experiencing to the distraction of imminent death, but it was more than that. The forms had been purposefully written in a confusing manner. They were, as the researchers put it, "an example of deliberately bad human engineering."

As soon as the men completed the forms, the pilot turned the plane around—*This is your captain speaking. Just kidding about that emergency.* Then he landed it safely.

The soldiers in the plane made a significantly greater amount of mistakes on the insurance forms than did a control group in a classroom on the ground who filled out the same forms, indicating the men felt stressed by the experience. But, disappointingly for the researchers who were hoping to produce a real scare, most of the men reported feeling merely "unsteady" during the incident. Perhaps filling out the forms actually calmed the men down by distracting them. Or maybe the plane needed to go into a nosedive to trigger a more dramatic reaction. A quarter of the subjects even figured out that the emergency was fake. Those with some flying experience realized something was fishy, but one soldier found a more direct clue: a note a subject in an earlier group had written on the back of an airsickness bag.

Unfazed, the researchers went back to the drawing board and devised three new situations. These all involved mishaps during a supposed "atomic-age warfare" exercise. Soldiers were driven out to remote rural outposts and dropped off alone. Their job, their commander told them, was to man a radio and notify headquarters should any planes fly overhead. Wearily, the soldiers prepared themselves for a long and boring day. But it didn't stay boring for long.

As the men sat sweating in the one-hundred-degree heat, an announcement suddenly crackled over the radio. Each man heard one of three warnings, depending on which experimental group he had been assigned to. He heard either that an accident with radioactive material had resulted in dangerous fallout over his area, that a forest fire was surrounding his position, or that misdirected artillery fire was incoming. "This is not a drill," headquarters emphasized. "Repeat. This is not a drill. Maneuvers have been cancelled. Radio in your position for immediate helicopter evacuation."

When the men tried to comply with the order they discovered, to their dismay, that their transmitters had chosen that moment to stop working. As if aware of the problem, headquarters gave another order: "Soldier, fix your transmitter and radio in your position." Fixing the radio was the task the experimenters had chosen to measure performance under pressure. On the outside of each radio was printed a wiring diagram that the men were supposed to consult to fix the instrument. However, the schematic was really "a visual pursuit subtest of the MacQuarrie Test of Mechanical Ability, revised to look like a wiring diagram."

Of the three situations, the radiation warning provoked, by far, the least reaction from the soldiers. Perhaps because the threat was invisible, the men acted as if there was little to fear. The researchers noted that "They tended to react as though the injury, if any, had already been suffered and that the only question remaining was that of establishing contact with the Command Post." In addition, many of the men appeared remarkably uninformed about the dangers of radiation. Evidently these young men had not paid enough attention in science class.

The forest fire elicited more interest. Upon hearing the warning, most of the men stood up to scan the horizon, at which point they saw billowing clouds of smoke about three hundred yards away—produced, unbeknownst to them, by smoke bombs. Two men panicked at the sight of the smoke and took off, but the majority remained calm and set to work on the radio. They later explained that they figured they could run away if the fire got any closer.

The clear first place in the fear-arousal contest went to the misdirected artillery fire. Seconds after the men heard the first warning on the radio—"Incoming artillery shells! Shells are landing outside the designated target area!"—a shell burst nearby. The soldiers threw themselves down on the ground and pulled on their flak jackets. They screamed into the transmitter, only to realize it wasn't working. A few continued to scream into it even after they knew it didn't work. Almost half of them took off running when a few more shells exploded, flagrantly disregarding the voice on the radio ordering them to remain at their post and repair the transmitter.

The lesson learned from these experiments was clear: If your goal is to arouse maximum fear, then subtlety is not a virtue. Loud, exploding bombs work best.

However, the larger goal of the experiments was to observe what psychological features characterized those who performed well under stress, in the hope that others could be trained to behave the same way. Here the results were far more tentative. The authors noted that, generally speaking, the more field experience and education a soldier had, the cooler he stayed under stress. They also noted that every top performer displayed the ability to "lose himself" in whatever task he was doing. These men were able to tune out the threat by "reducing imagery content centering around fear of harm or of physical injury."

Of course, those not in the military have slightly different priorities. Their goal is not to remain in position and continue to obey orders, but simply to survive. For which purpose, running at the first sign of radiation, forest fire, or incoming artillery shells, or screaming as a plane plunges to earth, still seem like compelling options.

Berkun, M. M., H. M. Bialek, R. P. Kern, & K. Yagi (1962). "Experimental Studies of Psychological Stress in Man." *Psychological Monographs: General and Applied* 76 (15, whole no. 534): 1–39.

Heartbeat At Death

October 31, 1938. 6:30 a.m. As John Deering walks to the room where he will be executed by firing squad, his face betrays no emotion. The sheriff reads the death warrant and Deering listens, casually sucking on a cigarette. The cigarette finished, he sits down in a chair positioned in front of the rock wall of the prison. A prison guard places a black hood over Deering's head and pins a target to his chest. Then prison physician Dr. Stephen H. Besley steps forward and attaches electronic sensors to Deering's wrists. Across the room, an electrocardiograph machine silently begins to record the hammering of the prisoner's heart.

Deering was not a typical death-row prisoner. When police picked him up on August 1, 1938, and charged him with the murder of Utah businessman Oliver Meredith, Deering readily admitted to the crime. He explained that he shot Meredith in cold blood while stealing the man's car. But Deering also expressed regret for what he had done and for the life he had led. He pleaded that the state kill him quickly "without all the red tape and rigamarole of courts." He got his wish. Only three months elapsed between his arrest and execution.

During the final weeks of his life, Deering attempted to be a model citizen. He spoke out on the need to provide children with more opportunities. "Build more athletic fields and gymnasiums," he wrote. "Give children more play facilities to keep their minds on wholesome activities. Give them the chance to develop that I never had."

In a gesture of atonement, Deering also willed his body to the University of Utah medical school and arranged for his eyes, following his death, to be frozen and flown to San Francisco, where a surgeon would attempt to use them to restore sight to a blind person. Finally, at the request of Dr. Besley, he agreed to participate in an experiment—the first of its kind—to have his heartbeat recorded during his execution. Dr. Besley believed the experiment would, besides satisfying morbid curiosity, reveal valuable information about the effect of fear on the heart, and how soon death occurs after the heart is wounded.

On the day of his execution, Deering walked stoically to the firing squad as his fellow prisoners banged on the bars of their cells and howled maniacally. He sat down in the chair and allowed Dr. Besley to attach the electrodes to his wrists.

The electrocardiogram immediately disclosed that, though Deering's face showed no emotion, his heart was beating like a jackhammer at 120 beats per minute, far higher than the resting average of 72 beats per minute.

The sheriff asked Deering whether he had any final words. His heartbeat momentarily fluttered higher. "I'd like to thank the warden for being so kind to me. Good-bye and good luck!" he replied. Then he murmured, "Okay, let it go."

The sheriff gave the order to fire. Deering's heartbeat raced up to 180 beats per minute. Then four bullets ripped into his chest, knocking him back into the chair. One bullet bore directly into the right side of his heart. For four seconds his heart spasmed. A moment later it spasmed again. Then the rhythm gradually declined until, 15.6 seconds after the first shot, Deering's heart stopped.

Although his heart no longer beat, his breathing continued for almost a minute as he twisted and squirmed in the chair. Finally, 134.4 seconds after his heart had stopped, he was pronounced dead. The time was 6:48 a.m.

The next day the grim experiment made headlines around the nation, sharing space with the mass panic caused by Orson Welles's Halloween-eve *War of the Worlds* radio broadcast. Dr. Besley offered the press a eulogy of sorts for Deering: "He put on a good front. The electrocardiograph film shows his bold demeanor hid the actual emotions pounding within him. He was scared to death."

Thanks to Dr. Besley's pioneering experiment, scientists can now say with certainty that the prospect of facing a firing squad causes a rapid heartbeat.

Midgley, L. (December 4, 1938). "You Can't Be Brave Facing Death." *Albuquerque Journal:* 19.

Dying on Acid

By the early 1960s the effects of LSD had been tested extensively. Cats, dogs, fish, mice, rats, baboons, chimpanzees, spiders, pigeons, and even, as we have seen, elephants, had all received the drug. It had been given to college students, prisoners, doctors, artists, government agents, soldiers, and tens of thousands of psychiatric patients. There weren't many groups left to try it on. Then Dr. Eric Kast of Chicago's Mount Sinai Hospital thought of what was, in hindsight, an obvious group. Not only might these people benefit from the drug, but they also didn't have much left to lose. They were the terminally ill.

Kast had observed that his terminally ill patients often became preoccupied with their imminent deaths. At a time when, ideally, they should have been striving to experience the remainder of their lives to the fullest and savoring time with friends and family, they instead grew depressed and withdrawn. "Interference," Kast wrote cautiously, "seems justified."

So Kast designed an experiment to study the effects of LSD on dying patients. He had no illusion that LSD could offer a cure, and he made sure all the test subjects knew that. Instead, he was interested in how LSD would alter the experience of facing death. LSD was reported to produce in recipients a sense of harmony with the surrounding universe. Kast described this as a "happy, oceanic feeling." Could LSD make the terminally ill more accepting of their fate and less fearful of approaching death?

Eighty patients took part in Kast's study. All had life expectancies measured in mere weeks or months. Kast gave them each one hundred micrograms of LSD delivered hypodermically. He then observed the drug's effects. If the subjects showed any sign of fear or disturbance—symptoms of a bad trip—he immediately administered an antipsychotic, chlorpromazine, that made them fall asleep. Most of the patients received the antipsychotic within eight to ten hours after being given the LSD. For the next three weeks, Kast interviewed and evaluated each patient daily. He paid careful attention to their moods, their attitudes toward life and death, and complaints of pain.

The results were encouraging. Of the eighty patients studied, seventy-two said they gained insight through the experience, fifty-eight found it pleasant, and sixty-eight (a full 85 percent) wanted to do it again.

The patients' attitudes toward life also showed definite signs of improvement. Before, during, and after the test Kast asked the patients to indicate which of three statements best approximated their current state of mind: (1) "I want to die, life has nothing to offer me"; (2) "I like to live, but it does not mean anything to me"; or (3) "Life is great, the concept of death does not frighten me." Before the test, most of them chose statement one; but while they were under the influence of LSD, number three became the favorite choice. Apparently, life seems great when you're high on acid, even if you're dying of cancer. Over the course of the following month, their moods evened out to number two.

The LSD did not directly block physical pain, but it did cause patients to focus less on their discomfort. Kast wrote that the drug seemed to reconcile them to their bodies. They felt the familiar aches and pains, but didn't worry about them as much.

One curious, less anticipated effect that Kast observed was the emergence of a sense of community and camaraderie among the participants. They would nudge one another and say, "Have you tried it? What do you think?" They acted like members of a secret club—not only special and privileged, but also somewhat superior to those around them who did not "know" the experience. They had become the cool in-crowd on the terminal ward.

All in all, Kast gave LSD a ringing endorsement:

> The results of this study seem to indicate that LSD is capable not only of improving the lot of pre-terminal patients by making them more responsive to their environment and family, but it also enhances their ability to appreciate the subtle and aesthetic nuances of experience. . . . Patients who had been listless and depressed were touched to tears by the discovery of a depth of feeling they had not thought themselves capable [of]. Although shortlived and transient, this happy state of affairs was a welcome change in their monotonous and isolated lives, and recollection of this experience days later often created similar elation.

Following Kast's study, a number of researchers conducted similar experiments. The Los Angeles psychiatrist Sidney Cohen supplied the drug to a handful of terminally ill patients —including, it is rumored, the author Aldous Huxley. (Huxley definitely did receive LSD on his deathbed, administered by his wife, Laura. The last words he ever wrote, scrawled on a piece of paper, were, "Try LSD 100 mm [sic] intramuscular." The only question is whether Cohen supplied the drug.) Walter Pahnke—famous for designing the so-called Miracle of Marsh Chapel experiment in 1962, in which he gave psilocybin to ten theology students as they participated in a Good Friday service—led a larger, more formal study of LSD and dying patients at Spring Grove State Hospital in Maryland during the late 1960s. Both Pahnke and Cohen reported results similar to those found by Kast.

However, funding for studies involving psychedelic drugs

dried up during the 1970s and '80s. Only recently have physicians begun actively lobbying to be able to pursue this line of research again, so they might be able to prescribe drugs such as LSD to the terminally ill.

In the meantime, medical workers interested in altering and improving the experience of dying have been searching for methods that do not involve the use of controlled substances. A practice called music thanatology—which involves playing music for dying people—has gained support. Popular deathbed musical choices include Gregorian chants or harp playing.

Music thanatology and LSD seem like naturally complementary forms of therapy, and maybe, if LSD is ever legalized again, their joint effects could be studied. Though the harp playing may have to go—a little Grateful Dead might be more appropriate.

Kast, E. (Summer 1966). "LSD and the Dying Patient." *Chicago Medical School Quarterly* 86: 80–87.

A Soul in the Balance

A man lies dying. He is motionless except for the twitching of a muscle in his face. A low rattle of phlegm accompanies each inhalation of breath. The bed he is lying on rests, in turn, on the large pan of a platform beam scale. Two doctors watch every quiver of the beam. Suddenly the whistling of the man's breath stops. The doctors look up from the scale and then glance at each other. "Is he dead?" one of the men whispers. As if in affirmation, the beam of the scale hits the lower bar with a distinct clang.

We speak metaphorically of people having a heavy soul, weighed down by grief or by the burden of years. Duncan MacDougall, a doctor who worked in Haverhill, Massachusetts, at the beginning of the twentieth century, took such talk literally. He reasoned that if there is such a thing as a soul, it must have a material basis. And if it has a material basis, then it must have weight. And if it has weight, then he should be able to weigh it.

But how exactly do you weigh a soul? MacDougall proposed a straightforward solution: Place a dying man on a scale and weigh him before and after death. Any unexplained difference between the two measurements would be, QED, the weight of the soul that had departed the body.

In 1900 MacDougall approached physicians at the nearby Cullis Free Home for Consumptives with his plan, and they gave him permission to conduct his experiment at their institution. All he had to do was wait for a patient to die.

Soon the doctors notified MacDougall that a tuberculosis patient was approaching his final hours. They moved the dying man's bed, with him in it, onto a Fairbanks scale, designed to weigh silk but adapted for its new purpose. Eagerly MacDougall calibrated the weights—accurate to one-tenth of an ounce. And then he waited. Occasionally he checked the patient. He listened for a heartbeat. He took the man's pulse. But the patient was in no hurry to die. As the time passed MacDougall kept a constant record of the scale's measurement, and discovered the man was losing weight at the rate of one ounce per hour.

Finally, after three hours and forty minutes, the patient exhaled for the final time. At the same instant, MacDougall wrote, "the beam end dropped with an audible stroke hitting against the lower limiting bar and remaining there with no rebound. The loss was ascertained to be three-fourths of an ounce."

MacDougall, in the best scientific fashion, tried to eliminate all other possibilities that might have accounted for this sudden weight loss. He ruled out a bowel movement or evacuation of the bladder, since the weight of these would have remained on the bed. Rapid evaporation of moisture from the patient's skin and lungs seemed unlikely. To exclude exhalation of air from the lungs, MacDougall lay down on the bed and exhaled as forcibly as he could while another doctor watched the scale. The beam of the scale didn't move. That left only one possibility, he concluded. The three-fourths of an ounce must have represented the weight of the soul.

Over the following months, MacDougall weighed five other patients with similar results—although he did admit to some problems that marred the reliability of his data. For instance, in one case people opposed to his work disrupted the experiment. MacDougall didn't elaborate on why they were opposed, but we can assume they were affiliated with one of the many religious groups hostile to his work, believing that the soul was an intangible object of faith and not something to be placed on a scale by scientists. Another time the patient died just as MacDougall was adjusting the scale.

You also get the sense that MacDougall was not a researcher who was about to let negative findings get in the way of his theories. He noted that in one patient the weight loss actually occurred a minute after death, but he breezily chalked this up to the patient's sluggish temperament. The man's soul, he asserted, was evidently as phlegmatic as the man's personality, causing it to linger in the body before departing.

When MacDougall subsequently conducted his experiment on fifteen dogs but found no loss of weight at death, he explained away this discrepancy also. He simply concluded that dogs do not have souls.

His work, when he published it in 1907, attracted the attention of the media, but didn't garner much respect from fellow scientists. The *Lancet* dismissed his findings, attributing them to "a peculiar bias on the part of his scales or on the part of the friends who assisted him."

Even those who shared MacDougall's interest in psychophysical phenomena treated his results with caution. Hereward Carrington, a member of the American Society for Psychical Research, advised that "the conditions attendant upon death are so little known, and the human organism is subject to such queer variations in weight, even when alive, that many and positive proofs will have to be forthcoming before [MacDougall's] interpretation of the facts, even though they themselves should be established, can be accepted by science."

However, Carrington, despite his caution, did propose a more rigorous version of MacDougall's experiment. He imagined using condemned prisoners as the subjects. An electric chair would be placed on a scale, and a glass hood fitted over the criminal's head, so that no moisture could escape from the lungs. "Then turn on the current, and at the instant of death watch for a loss of weight." No prison warden has ever taken Carrington up on this suggestion.

Since 1907, the mainstream scientific community has almost entirely ignored MacDougall's work, leaving it to the likes of Lewis Hollander, an Oregon sheep farmer, to continue this line of research. In 2001 Hollander reported the results of an experiment in which he weighed eleven plastic-wrapped sheep (and one goat) as they died. The plastic wrap was to contain "any voiding and fluid losses." Surprisingly, Hollander observed a brief weight gain of 18 to 780 grams at the moment of death. Where this extra weight came from, he didn't speculate. He published his study in the *Journal of Scientific Exploration*, a journal that appropriately describes itself as "committed to the rigorous study of unusual and unexplained phenomena."

Despite the lack of scientific interest in MacDougall's experiment, the idea that the soul has a weight has lived on in popular lore. The concept even found its way into the title of a Hollywood movie, *21 Grams*, starring Sean Penn and Benicio Del Toro. The movie's screenwriter converted three-quarters of an ounce (the soul-weight that MacDougall had measured) into metric to arrive at twenty-one grams. Of course, the movie didn't have anything to do with the experiment, but MacDougall would certainly still have been pleased by the publicity.

Surprisingly, the man who weighed other people's souls didn't bother to arrange to have his own soul weighed when he died, which he did in 1920, succumbing to liver cancer at the age of fifty-four. However, MacDougall remained interested in the process of dying right up to the end, closely observing his body as it succumbed to disease. According to his obituary in the local paper, he described his own death as "the most interesting he ever watched."

MacDougall, D. (April 1907). "Hypothesis Concerning Soul Substance Together with Experimental Evidence of the Existence of Such Substance." *American Medicine* 2 (4): 240–43.

The Day The World Didn't End

It is ten minutes to midnight. Fourteen people sit staring at a clock as the second hand creeps forward. They grip their overcoats tightly, ready to leave at any moment.

"Charles, do you remember the password?" a thin, middle-age woman sitting at the front of the group asks.

"Yes, Dorothy. We've practiced it a hundred times."

"Let's practice it just once more, to be sure."

Charles sighs, then nods. "Okay. At midnight, the spaceman will knock on the door. I will answer and ask, 'What is your question?'" He looks at Dorothy.

"He will reply, 'I am the porter,'" she says.

"And I will say, 'I am my own porter.'"

Dorothy nods with satisfaction. Silence falls on the room again as the group returns to its vigil, watching the minute hand approach midnight.

Six minutes pass. Dorothy shifts nervously in her seat. She clasps her hands together, looks upward as though in prayer, and says emphatically, "And not a plan has gone astray!" The others nod appreciatively.

The minute hand is only inches away from midnight. The tension in the room feels like a physical presence pressing down on everyone. The minute hand moves closer, closer, and finally slides over the number twelve. The clock begins to chime. Each note echoes in the room.

Everyone holds their breath as they strain to hear a noise at the door. But there's nothing.

Minutes pass. No one has knocked at the door. The members of the group look toward Dorothy questioningly. She stares downward, lost in her thoughts. Then, at last, she breaks the silence—"There has been a slight delay."

In late September 1954, American newspapers reported some bad news. In just three months, on the morning of December 21, a massive flood would create a vast inland sea stretching from the Arctic Circle to the Gulf of Mexico. Chicago, Detroit, and all the other towns and cities in the Midwest would be destroyed by tidal waves. Simultaneously, cataclysms would submerge the western coasts of the Americas, from Washington State to Chile. Similar disasters would devastate much of the rest of the world. Most of the planet's people were going to die.

What was the source of this dire prediction? A team of university researchers? A maverick scientist perhaps? No. It was a fifty-three-year-old Chicago grandmother named Dorothy Martin. She, in turn, was told of the impending holocaust by space aliens from the planet Clarion.

The media treated the prediction as a big joke, but it fascinated Leon Festinger, a young psychology professor at the University of Minnesota. Dorothy Martin obviously deeply believed her prediction, as did her small band of followers. They had risked public ridicule to warn the world of its approaching doom. But what was going to happen, he wondered, when the world didn't end? How would Martin's group deal with such a blow to their convictions? Festinger realized a natural experiment in the "disconfirmation of belief" was unfolding before his eyes. He resolved that the phenomenon be studied in person.

Festinger quickly put together a *Mission Impossible*–style team consisting of himself, two other social psychologists (Henry Riecken and Stanley Schachter), and a couple of graduate students. Their mission, which they all chose to accept, was to infiltrate Martin's group by posing as believers, to observe and record the actions of the group members in as much detail as possible, and to be there on December 21 when the world failed to end. They wanted to witness, first-hand, the group's reaction.

Festinger had a prediction of his own about how it all would turn out. He theorized that the dramatic disconfirmation wouldn't weaken the group's beliefs in the least. In fact, it would intensify them and prompt the group to make efforts to recruit more members. Why did Festinger predict this? Because he had been developing a theory he called "cognitive dissonance."

Festinger argued that people need their beliefs to be consistent and compatible. Incompatible beliefs (dissonant cognitions) cause psychological tension. For instance, if your belief system tells you the world should have ended, but it didn't, you'll need to resolve this discrepancy. A simple way to do so would be to discard your disproven beliefs. However, if you have already deeply committed yourself to those beliefs—for instance, if you have quit your job, left your spouse, and risked getting locked away in a mental asylum on account of your convictions—accepting that you were wrong might not be so easy. In such a case, it might paradoxically be easier to try to strengthen your belief by attempting to recruit other believers—because convincing someone else to share your ideas is like getting a vote of confidence. Suddenly your being right seems possible again. Festinger wrote, "If more and more people can be persuaded that the system of belief is correct, then clearly it must, after all, be correct."

The team of observers went to work infiltrating Martin's group. This required some creativity because Martin and her followers were—despite their message to the media—a quiet, reclusive bunch. They weren't seeking new recruits. So there was no application form the researchers could fill out. Instead, the observers approached the group individually with invented reasons for wanting to join that involved stories designed to appeal to the believers' philosophy. One grad student claimed she had dreamed of a terrible flood, and then had seen the prediction in the paper. Another observer told of meeting a mysterious, space-alien-like stranger in the desert. The deception worked, and soon all the observers were warmly accepted into the group.

The only problem? By showing up en masse with all these wild stories, the observers powerfully reinforced the groups' beliefs. Martin decided the space aliens were sending people to her for instruction and committed herself even more fervently to her beliefs. So, instead of merely observing, the researchers had, from the start, altered the course of events through their presence.

The researchers established a home base in a hotel room a few blocks from Martin's house and took turns hanging out with the group. They took notes about ongoing events whenever they could, sometimes by excusing themselves to go to the bathroom (though not often enough to attract attention) or by stepping outside and frantically scribbling down observations in the dark. Or they waited until they got back to the hotel and immediately dictated everything they could remember into a tape recorder.

The biggest challenge the researchers faced was maintaining a neutral role in the group. Martin's ideology, they wrote, "aroused constant incredulity." Often they wanted to shout, *What are you guys thinking!* Instead, they had to smile and go along with everything.

Martin's belief system was an eclectic mix of Christianity, New Age mysticism, and pulp zine science fiction. She claimed to be receiving messages from Clarion, a planet where bodies automatically adjusted to the outside temperature, people ate snowflakes, and no one ever died. Messages came to her through the spirit of Sananda, who apparently was Jesus Christ going by a different name. She received the messages by going into a trancelike state and allowing the aliens to guide her hand as she wrote down words on a piece of paper —a process called automatic writing.

Martin told her followers floods would destroy much of Earth on December 21, but space aliens would descend in a ship and rescue them, the true believers, before then. She set the time of the rescue at midnight on the 21st, but she expected the ship might show up early. Therefore she was constantly sending her followers out on "saucer watch," scanning the skies for stray spacecraft. She urged everyone to remain in a state of readiness by keeping metal, such as zippers or belt buckles, off their persons. On a spaceship, contact with metal would cause severe burns. Martin never explained why this was so but assured everyone it had to do with advanced alien technology.

The researchers expressed amazement at what easy marks the group made themselves for the pranksters who, as December 21 approached, targeted Martin with increasing frequency. She and her followers would almost always take the bait, no matter how ridiculous or obvious the trick was. One time a young man called claiming to be "Captain Video from Outer Space." He told them a spaceship would pick them up at noon. Obediently, the group members trooped outside to wait in the snow for the ship. Another time some boys called up saying they had a flood in their bathroom that they wanted the group to see. Martin, believing the boys were spacemen in disguise, had everyone go over. About the only invitation they didn't accept was one to an "end of the world cocktail party" a reporter suggested they attend.

The tension steadily built until the night of December 20, when the entire group gathered in Martin's living room to wait for midnight. They waited and waited, but no spacemen appeared. At 12:30 a.m. there was a knock at the door, which caused a flurry of excitement. A member went to answer the door as Martin called after him to remember the secret password, but he returned seconds later with the news that it was just some boys playing games. Finally, at 2:30 a.m., Martin announced she had received another message from Sananda. He wasn't apologizing for blowing them off. Instead, the important message he had beamed all the way from the planet Clarion was that he wanted them to take a coffee break.

As the members of the group milled around, drinking their coffee, the observers pressed them for their reactions to the nonarrival of the spaceship. Many of Martin's followers had staked everything on the assumption that spacemen were going to spirit them away. They had quit their jobs and spent all their money. What were they going to do now that they were stuck on Earth? But no one felt like talking. There was a mood of uncomfortable tension. Some members walked around blankly, seemingly disillusioned. Confusion reigned. They were all waiting for Martin to explain why nothing had happened, and at 4:45 a.m. she finally did exactly that. She announced the receipt of a new message from Clarion:

> For from the mouth of death have ye been delivered and at no time has there been such a force loosed upon the Earth. Not since the beginning of time upon this Earth has there been such a force of Good and light as now

floods this room and that which has been loosed within
this room now floods the entire Earth.

What did it mean? It meant, she explained, that they had
saved the world! Their devout belief had averted the cat-
astrophe. That's why the spaceship hadn't come. Soon, a
second message arrived. The spacemen wanted them to
spread the "Christmas Message" of joy and salvation to the
entire world. Everyone needed to know of the glorious
redemption.

It was just as Festinger had predicted. The stunning disconfirmation of the prediction hadn't dented the followers'
beliefs at all. Instead, it made their convictions stronger and
mobilized the group to seek out new members. Whereas
before, Martin and her followers had shunned publicity, the
morning of December 21 found them on the phone to
reporters, drumming up media attention. Martin made audio-
tapes of her messages available. She issued a press release.
Later, the entire group sang Christmas carols on the lawn,
both to spread the message of joy to their neighbors and in a
last-ditch attempt to attract a spaceship.

However, despite great efforts, the group didn't attract a
single convert. It turned out they were lousy at proselytizing.
The researchers wrote:

> For about a week they were headline news throughout the
> nation. Their ideas were not without popular appeal, and
> they received hundreds of visitors, telephone calls, and let-
> ters from seriously interested citizens, as well as offers of
> money (which they invariably refused). Events conspired
> to offer them a truly magnificent opportunity to grow in
> numbers. Had they been more effective, disconfirmation
> might have portended the beginning, not the end.

Of course, that a significant percentage of the members were planted stooges, cynically observing all that went on as part of a science experiment, somewhat undermined the group's effectiveness.

Festinger's research offers a gloomy lesson about the resiliency of beliefs. Have you ever gotten into an argument with someone who wouldn't change his mind no matter what facts, evidence, or logic you presented him with? The case of Dorothy Martin and her followers suggests you might as well give up the effort, because beliefs can easily survive being disproven—and can in fact become stronger as a result. Lurking in the background of Festinger's thesis is the idea that disconfirmation may have been the triggering event responsible for the spread of many religions.

So what was the aftermath of the failure of the world to end in 1954? Festinger, Riecken, and Schachter wrote an account of their research that they titled *When Prophecy Fails*. To protect Martin's privacy and to shield themselves from lawsuits, they referred to Martin as Marian Keech and set all the events in a fictitious Lake City (rather than Chicago). However, it was never much of a secret that Dorothy Martin was the subject of their study. After all, the Christmas Message that Marian Keech delivers in *When Prophecy Fails* is the same, word for word, as Dorothy Martin's Christmas Message, which appeared in many newspapers in December 1954.

Martin carried on her career as a New Age prophet. She changed her name to Sister Thedra and traveled to South America, where she established a small religious center called the Abbey of the Seven Rays. She continued to predict a coming time of floods, when a new Atlantis would rise from the oceans, but she grew less specific about the date when all this would happen. Eventually she returned to the United

States, where she died in 1988. Or perhaps, we should say, her spaceship finally arrived.

———
Festinger, L., H. W. Riecken, & S. Schachter (1956). *When Prophecy Fails: A Social and Psychological Study of a Modern Group that Predicted the Destruction of the World*. New York: Harper Torchbooks.

The Last Survivor

It's the end of the world. The bombs have fallen. The mushroom clouds have bloomed and faded on the horizon. Finally, nothing remains of human civilization except a charred, radioactive ruin. But one creature survives. It crawls from the smoking rubble, clambers to the top of the wreckage, and waves its antennae in victory. It is a cockroach.

We've all heard the claim that in the event of a nuclear war, cockroaches will be the only survivors. But where does this idea come from? Do people say this just because the bugs look tough enough to survive anything, or has someone actually irradiated a bunch of cockroaches to measure precisely how many rads they can withstand?

By now you can probably guess that, yes, someone has irradiated cockroaches. In 1959, at the Quartermaster Research and Engineering Center in Natick, Massachusetts, the Whartons (D. R. A. and Martha) performed what remains the one definitive experiment on this question. They filled polyethylene bags with twenty to twenty-five cockroaches (*Periplaneta americana*), inserted a breathing tube into the bag so the little guys had some air, and then placed the bags on a conveyor belt that ran through a two-MeV Van de Graaff electron accelerator. Different groups of roaches were exposed to varying amounts of radiation.

Subsequently, the Whartons placed each cockroach in a beaker, gave it some dog food—apparently roaches love the stuff—and waited to see how long it would live.

Surprisingly, given the reputation of roaches, the critters didn't fare very well. One thousand rads will kill a human. The same amount made the roaches sterile. So even if they do survive the bomb, they won't be breeding much. Ten thousand rads stunned them. At 40,000 rads they died.

These amounts are far more than humans could survive, but the subjects' response was not enough to guarantee roaches will rule a postapocalyptic planet Earth. So the legend of the radiation-proof roach is just that—a legend.

The true lord of radiation, it turns out, is the parasitoid wasp *Habro bracon*. It takes an unbelievable 180,000 rads to be sure of killing it, as the researchers R. L. Sullivan and D. S. Grosch discovered in 1953. Which means, to paraphrase T. S. Eliot, the world will end not with the bang of bombs or the hiss of cockroaches, but with the buzzing of wasps.

———

Wharton, D. R. A., & M. L. Wharton (1959). "The Effect of Radiation on the Longevity of the Cockroach, *Periplaneta americana*, as Affected by Dose, Age, Sex and Food Intake." *Radiation Research* 11: 600–15.

ACKNOWLEDGMENTS

Thanks to my editor, Stacia Decker, for providing me with the opportunity to write this book and for the numerous improvements she made to it. Thanks also to my agent and fellow *Lost* fan, Alička Pistek.

Sally Richards deserves special credit for, week after week, offering her thoughts and comments on the manuscript—as well as for keeping me on target to finish on time.

The love and support of my family and friends kept me going during the months of writing. Beverley—I absolutely couldn't have finished without you. Mom and Dad—I'm incredibly lucky to have you as parents. Ted—once again, you came through with the coffee breaks. Charlie—how could I even begin to list all the things I should thank you for? Kirsten, Ben, Astrid, and Pippa—I wish I could see you guys more often, but at least I knew you were cheering me on from Malawi. Boo—as a spoiled little cat, I'm sure you know how important you are.

Flora Streater greatly helped me by keeping the Museum of Hoaxes operating while I took a leave of absence to work on the book. And thanks to all the other site regulars, especially the gang from the Edinburgh get-together—Annette Hudson (Nettie), Rowenna Streater (Madmouse), Sarah Kirkham (Smerk), Amber Belken (Tru), and William Wilhite (Charybdis)—for keeping the site active while I went off chasing elephants.

REFERENCES

References are listed according to the sections in which they appear in the book. Those already footnoted in the text are not repeated here. For some sections there are no additional references beyond the one in the text.

One: Frankenstein's Lab

THE BODY ELECTRIC

Farrar, W. V. (1973). "Andrew Ure, F. R. S., and the Philosophy of Manufactures." *Notes and Records of the Royal Society of London* 27 (2): 299–324.

London Times (January 22, 1803), page 3, column D.

London Times (February 15, 1803), page 3, column C.

Morus, I. R. (1998). *Frankenstein's Children: Electricity, Exhibition, and Experiment in Early-Nineteenth-Century London.* New Haven, CT: Princeton University Press. 125–52.

Pera, M. (1992). *The Ambiguous Frog: The Galvani-Volta Controversy on Animal Electricity.* Translated by Jonathan Mandelbaum. New Haven, CT: Princeton University Press.

ZOMBIE KITTEN

Finger, S., & M. B. Law (1998). "Karl August Weinhold and his 'Science' in the era of Mary Shelley's Frankenstein: Experiments on Electricity and the Restoration of Life." *Journal of the History of Medicine* 53: 161–80.

THE ELECTRICAL ACARI

Crosse, C. (1857). *Memorials, scientific and literary of Andrew Crosse, electrician.* Longman: 353–60.

Haining, P. (1979). *The Man Who Was Frankenstein.* London: Frederick Muller.

Miller, S. L. (1953). "A Production of Amino Acids under Possible Primitive Earth Conditions." *Science* 117 (3046): 528–29.

Secord, J. A. (1988). "Extraordinary Experiment: Electricity and the Creation of Life in Victorian England." In *The Uses of Experiment*, Gooding, D., T. Pinch, & S. Schaffer. Cambridge: Cambridge University Press. 337–83.

SEVERED HEADS – AN ABBREVIATED HISTORY

Brown-Séquard, É. (1858). L'encephale, après avoir completement perdu ses fonctions et ses propriétes vitales peut les recouvrer sous l'influence de sang charge d'oxygene. *Journal de la physiologie de l'homme et des animaux*. Paris: Tome Premier. 117–22.

Brukhonenko, S. (1929). Expériences avec la tête isolée du chien II: Résultats des experiences. *Journal de physiologie et de pathologie générale* 27 (1): 65–79.

Experiments in the Revival of Organisms (1940). Soviet Film Agency. Viewable at: http://www.archive.org/details/Experime1940.

Hecht, J. M. (1997). French Scientific Materialism and the Liturgy of Death: The Invention of a Secular Version of Catholic Last Rites (1876–1914). *French Historical Studies* 20 (4): 703–35.

Loye, P. (1888). *La mort par la décapitation*. Bureaux du Progres medical. Paris.

"An Outrage Against Humanity" (January 5, 1885). *Galveston Daily News*: 3.

Shaw, G. B. (March 17, 1929). Shaw will sich köpfen lassen, wenn . . . Ein Privatbrief des Dichters über ein neues, abschreckendes Tierexperiment. *Berliner Tageblatt*: 1.

HUMAN-APE HYBRID

Patterson, N., et al. (2006). "Genetic evidence for complex speciation of humans and chimpanzees." *Nature* 441 (7097): 1103–8.

THE MAN WHO CHEATED DEATH

"Cornish Readies Life Machine" (May 15, 1947). *Oakland Tribune*: 16.

"Dr. Cornish, Chemist, Dies at 59" (March 6, 1963). *Oakland Tribune*: 1–2.

Ford, J. E. (February 1935). "Can Science Raise the Dead?" *Popular Science Monthly* 126 (2): 11–13, 108.

Life Returns (1935). Scienart Pictures. Available on DVD from Alpha Video: http://www.oldies.com.

Shuster, E. (March 16, 1934). "Life-Giving Fluid Is Injected into Dead Dog's Veins and Breath Breathed into Lungs to Restore Life to Him." *The Burlington (N.C.) Daily Times-News*: 8.

THE TWO-HEADED DOGS OF DR DEMIKHOV

Mosby, A. (April 26, 1959). "Two-Headed Russian Dog Displayed for Reporters." *Nevada State Journal*: 8.

FRANKEN-MONKEY

Fallaci, O. (November 28, 1967). "The Dead Body & the Living Brain." *Look*: 99–108.

White, R. J., et al. (1996). "The isolation and transplantation of the brain. An historical perspective emphasizing the surgical solutions to the design of these classical models." *Neurological Research* 18 (3): 194–203.

Two: Sensorama

THE MOCK-TICKLE MACHINE

Harris, C. R. (1999). "The Mystery of Ticklish Laughter." *American Scientist* 87 (4): 344–48.

TOUCHING STRANGERS

Hornik, J. (1992). "Tactile Stimulation and Consumer Response." *The Journal of Consumer Research* 19 (3): 449–58.

Ovesen, L. (2004). "The Midas touch and other tipping stunts." *European Journal of Cancer Prevention* 13: 465–66.

Silverthorne, C. (1972). "The Effects of Tactile Stimulation on Visual Experience." *Journal of Social Psychology* 88: 153–54.

Stephen, R., & R. L. Zweigenhaft (1986). "The Effect of Tipping of a Waitress Touching Male and Female Customers." *Journal of Social Psychology* 126: 141–42.

WHAT'S THE DIFFERENCE?

Sage, A. (January 14, 2002). "Cheeky little test exposes wine 'experts' as weak and flat." *Times* (London).

Zwerdling, D. (August 2004). "Shattered Myths." *Gourmet*: 72–74, 126.

COKE VS. PEPSI

Thompson, C. (October 26, 2003). "There's a Sucker Born in Every Medial Prefrontal Cortex." *New York Times Magazine*: 54–57.

SYCHRONOUS MENSTRUATION (THE SCENT OF A WOMAN)

Horn, M. (1999). *Rebels in White Gloves: Coming of Age with Hillary's Class, Wellesley '69*. New York: Times Books. 123–32.

Stern, K., & M. K. McClintock (1998). "Regulation of Ovulation by Human Pheromones." *Nature* 392: 177–79.

Wright, K. (1994). "The Sniff of Legend; Human Pheromones? Chemical Sex Attractants? And a Sixth Sense Organ in the Nose? What Are We, Animals?" *Discover* 15 (4): 60–68.

THE SMELL OF MONEY

Kleinfield, N. R. (October 25, 1992). "The Smell of Money." *New York Times:* VI, 8.

Trivedi, B. (2006). "The Hard Smell." *New Scientist* 192 (2582): 36–39.

SMELL ILLUSIONS

De Araujo, I. E., E. T. Rolls, M. I. Velazco, C. Margot, & I. Cayeux (2005). "Cognitive Modulation of Olfactory Processing." *Neuron* 46 (4): 671–79.

Slosson, E. E. (1899). "A Lecture Experiment in Hallucinations." *Psychological Review* 6: 407–8.

THE INVISIBLE GORILLA

Simons, D. J., & D. T. Levin (1998). "Failure to Detect Changes to People During a Real-World Interaction." *Psychonomic Bulletin & Review* 5 (4): 644–49.

THE MOZART EFFECT

Bangerter, A., & C. Heath (2004). "The Mozart Effect: Tracking the Evolution of a Scientific Legend." *British Journal of Social Psychology* 43: 605–23.

Chabris, C. F. (1999). "Prelude or Requiem for the 'Mozart Effect'?" *Nature* 400 (6747): 826–27.

Hetland, L. (2000). "Listening to Music Enhances Spatial-Temporal Reasoning: Evidence for the 'Mozart Effect.'" *Journal of Aesthetic Education* 34 (3/4): 105–48.

THE ACOUSTICS OF COCKTAIL PARTIES

Lebo, C. P., K. S. Oliphant, & J. Garrett (1967). "Acoustic Trauma from Rock-and-Roll Music." *California Medicine* 107 (5): 378–80.

MacLean, W. R. (1959). "On the Acoustics of Cocktail Parties." *The Journal of the Acoustical Society of America* 31 (1): 79–80.

Three: Total Recall

ELECTRIC RECALL

Loftus, E. F., & G. R. Loftus (1980). "On the Permanence of Stored Information in the Human Brain." *American Psychologist* 35 (5): 409–20.

Penfield, W. (1958). "Some Mechanisms of Consciousness Discovered during Electrical Stimulation of the Brain." *Proceedings of the National Academy of Sciences of the United States of America* 44 (2): 51–66.

Valenstein, E. S. (1973). *Brain Control: A Critical Examination of Brain Stimulation and Psychosurgery.* New York: John Wiley & Sons. 108–14.

ELEPHANTS NEVER FORGET

Markowitz, H., M. Schmidt, L. Nadal, & L. Squier (1975). "Do Elephants Ever Forget?" *Journal of Applied Behavior Analysis* 8 (3): 333–35.

Rensch, B. (1956). "Increase of Learning Capability with Increase of Brain-Size." *The American Naturalist* 90 (851): 81–95.

THE MEMORY SKILLS OF COCKTAIL WAITRESSES

Ingram, J. (2000). *The Barmaid's Brain: And Other Strange Tales from Science.* New York: W. H. Freeman and Company.

UNDERWATER MEMORY

Godden, D., & A. Baddeley (1980). "When Does Context Influence Recognition Memory?" *British Journal of Psychology* 71: 99–104.

Koens, F., O. T. J. T. Cate, & E. J. F. M. Custers (2003). "Context-Dependent Memory in a Meaningful Environment for Medical Education: In the Classroom and at the Bedside." *Advances in Health Sciences Education* 8: 155–65.

EDIBLE MEMORY

Bird, J. (March 28, 1964). "The Worm Learns." *Saturday Evening Post:* 66–67.

Gratzer, W. (2000). *The Undergrowth of Science: Delusion, Self-Deception and Human Frailty.* New York: Oxford University Press. 57–64.

Rilling, M. (1996). "The Mystery of the Vanished Citations: James McConnell's Forgotten 1960s Quest for Planarian Learning, a Biochemical Engram, and Celebrity." *American Psychologist* 51 (6): 589–98.

Travis, G. D. L. (1981). "Replicating Replication? Aspects of the Social Construction of Learning in Planarian Worms." *Social Studies of Science* 11 (1): 11–32.

Ungar, G., L. Galvan, & G. Chapouthier (1972). "Evidence for Chemical Coding of Color Discrimination in Goldfish Brain." *Experientia* 28 (9): 1026–27.

BENEFICIAL BRAINWASHING

Cameron, D. E. (1960). "Production of Differential Amnesia as a Factor in the Treatment of Schizophrenia." *Comprehensive Psychiatry* 1: 26–34.

Collins, A. (1997). *In the sleep room: The Story of the CIA Brainwashing Experiments in Canada.* Toronto: Key Porter Books.

Gillmor, D. (1987). *I Swear by Apollo: Dr. Ewen Cameron and the CIA-Brainwashing Experiments.* Montreal: Eden Press.

Marks, J. (1979). *The Search for the Manchurian Candidate: The CIA and Mind Control.* New York: Times Books. Chapter 8.

THE WHITE BEAR

Wegner, D. M. (1989). *White Bears and Other Unwanted Thoughts: An Exploration of Suppression, Obsession, and the Psychology of Mental Control.* New York: Viking.

Wegner, D. M., & D. J. Schneider (2003). "The White Bear Story." *Psychological Inquiry* 14 (3&4): 326–29.

LOST IN THE MALL

Loftus, E. F., & K. Ketcham (1996). *The Myth of Repressed Memory: False Memories and Allegations of Sexual Abuse.* New York: St. Martin's Griffin.

Neimark, J. (1996). "The Diva of Disclosure." *Psychology Today* 29 (1): 48–52, 78, 80.

Four: Bedtime Stories

Martin, P. (2004). *Counting Sheep: The Science and Pleasures of Sleep and Dreams.* New York: St. Martin's Press.

SLEEP LEARNING

"Deeper . . . Deeper . . . Dee . . ." (March 20, 1950). *Time:* 77.

Elliott, C. R. (1947). "An Experimental Study of the Retention of Auditory Material Presented During Sleep." Unpublished master's thesis, University of North Carolina.

Emmons, W. H., & C. W. Simon (1956). "The Non-recall of Material Presented during Sleep." *The American Journal of Psychology* 69 (1): 76–81.

Fox, B. H., & J. S. Robbin (1952). "The Retention of Material Presented during Sleep." *Journal of Experimental Psychology* 43: 75–79.

"He Teaches Frogs to Lose Hangups." (December 17, 1972). *The Daily Review* (Hayward, Calif.): 14.

"Learning while you sleep method eases home work." (September 6, 1955). *Albuquerque Journal*: 26.

ELEVEN DAYS AWAKE

De Manaceine, M. (1894). "Quelques observations expérimentales sur l'influence de l'insomnie absolue." *Archives Italiennes de biologie* 21: 322–25.

Dement, W. C. (1974). *Some Must Watch While Some Must Sleep*. San Francisco: W. H. Freeman.

Patrick, G. T. W., & J. A. Gilbert (1896). "On the Effects of Loss of Sleep." *The Psychological Review* 3 (5): 469–83.

LET SLEEPING CATS HUNT

Brown, C. (February 2, 2003). "The Man Who Mistook His Wife for a Deer." *New York Times Magazine*: 34–41, 53, 72, 79, 82, 83.

Jouvet, M. (1967). "The States of Sleep." *Scientific American* 216 (2): 62–72.

Hendricks, J. C., A. R. Morrison, & G. L. Mann (1982). "Different behaviors during paradoxical sleep without atonia depend on pontine lesion site." *Brain Research* 239:81–105.

Henley, K., & A. R. Morrison (1974). "A re-evaluation of the effects of lesions of the pontine tegmentum and locus coeruleus on phenomena of paradoxical sleep in the cat." *Acta Neurobiologiae Experimentalis* 34: 215–32.

WHAT DREAMS MAY COME

"Sweet Dreams Are Made of Cheese." (September 25, 2005). British Cheese Board press release. Available online at: http://www.cheeseboard.co.uk/news.cfm?page_id=240.

Tauber, E. S., H. P. Roffwarg, & J. Herman (1968). "The effects of longstanding perceptual alterations on the hallucinatory content of dreams." *Psychophysiology* 5: 219.

Five: Animal Tales

ELEPHANTS ON ACID

Conley, C. (August 4, 1962). "Shot of Drug Kills Tusko." *Daily Oklahoman:* 1–2.

"Elephant Dies from New Drug" (August 5, 1962). *Appleton Post-Crescent:* A2.

"Fatal Research: Drug Kills Elephant Guinea Pig" (August 4, 1962). *Long Beach Press-Telegram:* B12.

Harwood, P. D. (1963). "Therapeutic Dosage in Small and Large Mammals." *Science* 139 (3555): 684–85.

Koella, W. P., R. F. Beaulieu, & J. R. Bergen (1964). "Stereotyped behavior and cyclic changes in response produced by LSD." *International Journal of Neuropharmacology* 3: 397–403.

Lemov, R. (2005). *World as Laboratory: Mice, Mazes, and Men.* New York: Hill and Wang. Chapter 10.

"LSD Related Death of an Elephant" (August 16, 2002). Erowid. Available online at: http://www.erowid.org/chemicals/lsd/lsd_history4.shtml.

"The Maestro of 'Mind-Control' Continues to Haunt America." *Freedom Magazine.* Available online at: http://www.freedommag.org/english/la/issue02/page12.htm.

Siegel, R. K. (1984). "LSD-Induced Effects in Elephants: Comparisons with Musth Behavior." *Bulletin of the Psychonomic Society* 22 (1): 53–56.

Siegel, R. K., & M. E. Jarvik (1975). "Drug-Induced Hallucinations in Animals and Man." In Siegel, R. K., & L. J. West, eds. *Hallucinations.* New York: John Wiley & Sons. 81–161.

Witt, P. N., C. F. Reed, & D. B. Peakall (1968). *A Spider's Web: Problems in Regulatory Biology.* New York: Springer-Verlag.

RACING ROACHES

Rajecki, D. W., W. Ickes, C. Corcoran, & K. Lenerz (1977). "Social Facilitation of Human Performance: Mere Presence Effects." *Journal of Social Psychology* 102: 297–310.

Worringham, C. J., & D. M. Messick (1983). "Social Facilitation of Running: An Unobtrusive Study." *Journal of Social Psychology* 121: 23–29.

EYEING AN UNGULATE

Ellsworth, P. C., J. M. Carlsmith, & A. Henson (1972). "The Stare as a

Stimulus to Flight in Human Subjects: A Series of Field Experiments." *Journal of Personality and Social Psychology* 21 (3): 302–11.

HORNY TURKEYS AND HYPERSEXUAL CATS

Carbaugh, B. T., M. W. Schein, & E. B. Hale (1962). "Effects of Morphological Variations of Chicken Models on Sexual Responses of Cocks." *Animal Behaviour* 10: 235–38.

Davis, K. (2001). *More Than a Meal: The Turkey in History, Myth, Ritual, and Reality.* New York: Lantern Books.

Schreiner, L., & A. Kling (1953). "Behavioral Changes Following Rhinencephalic Injury in Cat." *Journal of Neurophysiology* 16: 643–59.

Stimuli Releasing Sexual Behavior of Domestic Turkeys. (1958). Produced by M. W. Schein & E. B. Hale. Available from Penn State Media Sales.

THE BRAIN SURGEON AND THE BULL

Horgan, J. (October 2005). "The Forgotten Era of Brain Chips." *Scientific American* 293 (4): 66–73.

Osmundsen, J. A. (May 17, 1965). "'Matador' with a Radio Stops Wired Bull." *New York Times:* 1, 20.

Valenstein, E. S. (1973). *Brain Control: A Critical Examination of Brain Stimulation and Psychosurgery.* New York: John Wiley & Sons. 99.

Six: Mating Behavior

LOVE AT LAST CALL

Madey, S. F., M. Simo, D. Dillworth, D. Kemper, A. Toczynski, & A. Perelle (1996). "They Do Get More Attractive at Closing Time, but Only When You Are Not in a Relationship." *Basic and Applied Social Psychology* 18 (4): 387–93.

Nida, S. A., & J. Koon (1983). "They get better looking at closing time around here, too." *Psychological Reports* 52: 657–58.

Sprecher, S., J. DeLamater, N. Neuman, M. Neuman, P. Kahn, D. Orbuch, & K. McKinney (1984). "Asking Questions in Bars: The Girls (and Boys) May Not Get Prettier at Closing Time and Other Interesting Results." *Personality and Social Psychology Bulletin* 10: 482–88.

THE GAY DETECTOR

Clarke, S. (1999). "Justifying deception in social science research." *Journal of Applied Philosophy* 16 (2): 151–66.

HEART RATE DURING INTERCOURSE

Bartlett, R. G. (1956). "Physiologic responses during coitus." *Journal of Applied Physiology* 9: 469–72.

PUSHING THE PLEASURE BUTTON

Heath, R. G. (1972). "Pleasure and Brain Activity in Man." *The Journal of Nervous and Mental Disease* 154 (1): 3–18.

Olds, J., & P. Milner (1954). "Positive Reinforcement Produced by Electrical Stimulation of Septal Area and Other Regions of Rat Brain." *Journal of Comparative and Physiological Psychology* 47: 419–27.

VOULEZ-VOUS COUCHEZ AVEC MOI (CE SOIR)?

Clark, R. D. (1990). "The Impact of AIDS on Gender Differences in Willingness to Engage in Casual Sex." *Journal of Applied Social Psychology* 20 (9): 771–82.

Clark, R. D., & E. Hatfield (2003). "Love in the Afternoon." *Psychological Inquiry* 14 (3&4): 227–31.

THE PENIS IMAGINED AS A SPERM-SHOVELLING SCOOP

Gallup, G. G., & R. L. Burch (2004). "Semen Displacement as a Sperm Competition Strategy in Humans." *Evolutionary Psychology* 2: 12–23.

Gallup, G. G., R. L. Burch, & S. M. Platek (2002). "Does Semen Have Antidepressant Properties?" *Archives of Sexual Behavior* 31 (3): 289–93.

Seven: Oh, Baby!

Sulek, Antoni. (1989). "The Experiment of Psammetichus: Fact, Fiction, and Model to Follow." *Journal of the History of Ideas* 50 (4): 645–51.

LITTLE ALBERT AND THE RAT

Benjamin, L. T., J. L. Whitaker, R. M. Ramsey, & D. R. Zeve (2007). "John B. Watson's Alleged Sex Research: An Appraisal of the Evidence." *American Psychologist* 62 (2): 131–39.

Cornwell, D., & S. Hobbs (1976). "The strange saga of Little Albert." *New Society.* March 18: 602–4.

Harris, Ben. (February 1979). "Whatever Happened to Little Albert?" *American Psychologist* 34 (2): 151–60.

Magoun, H. W. (1981). "John B. Watson and the Study of Human Sexual Behavior." *Journal of Sex Research* 17 (4): 368–78.

Watson, J. B., & R. R. Watson (December 1921). "Studies in Infant Psychology." *Scientific Monthly* 13 (6): 493–515.

SELF-SELECTION OF DIET BY INFANTS

Davis, C. M. (1935). "Self-Selection of Food by Children." *The American Journal of Nursing* 35 (5): 403–10.

Davis, C. M. (1939). "Results of the self-selection of diets by young children." *Canadian Medical Association Journal* 41 (3): 257–61.

Munro, N. (1966). "A Review of the 1928 Research by Clara Davis." *Journal of Home Economics* 58 (8): 655–58.

A GIRL NAMED GUA

Benjamin, L. T., Jr., & D. Bruce (1982). "From Bottle-Fed Chimp to Bottlenose Dolphin: A Contemporary Appraisal of Winthrop Kellogg." *The Psychological Record* 32: 461–82.

BABY IN A BOX

Benjamin, L. T., & E. Nielsen-Gammon (1999). B. F. Skinner and Psychotechnology: The Case of the Heir Conditioner. *Review of General Psychology* 3 (3): 155–67.

Capshew, J. H. (October 1993). "Engineering Behavior: Project Pigeon, World War II, and the Conditioning of B. F. Skinner." *Technology and Culture* 34 (4): 835–57.

Skinner, B. F. (March 1979). "My Experience with the Baby-Tender." *Psychology Today:* 29–31, 34, 37–38, 40.

THE NEW MOTHER

Blum, D. (2002). *Love at Goon Park: Harry Harlow and the Science of Affection.* New York: Perseus Publishing.

Harlow, H. F., & S. J. Suomi (1970). "Nature of Love—Simplified." *American Psychologist* 25 (2): 161–68.

Mason, W. A. (1978). "Social Experience and Primate Cognitive Development." In Burghardt, G. M., & M. Bekoff, eds. *The Development of Behavior: Comparative and Evolutionary Aspects.* New York: Garland Publishing. 233–51.

THE ULTIMATE BABY MOVIE

Wright, S. H. (May 17, 2006). "Media Lab Project Explores Language Acquisition." *MIT Tech Talk* 50 (27): 4.

Eight: Toilet Reading

Kira, A. (1976). *The Bathroom.* New York: The Viking Press.

SPACE INVADERS IN THE LOO
Koocher, G. P. (1977). "Bathroom Behavior and Human Dignity."
Journal of Personality and Social Psychology 35 (2): 120–21.

Middlemist, R. D., E. S. Knowles, & C. F. Matter (1977). "What to Do
and What to Report: A Reply to Koocher." *Journal of Personality and
Social Psychology* 35 (2): 122–24.

FART-OLOGY
Tomlin, J., C. Lowis, & N. W. Read (1991). "Investigation of normal
flatus production in healthy volunteers." *Gut* 32: 665–69.

Nine: Making Mr. Hyde

SHOCKING OBEDIENCE
Blass, T. (2004). *The Man Who Shocked the World: The Life and Legacy
of Stanley Milgram.* New York: Basic Books.

Masserman, J. H., S. Wechkin, & W. Terris (December 1964).
"'Altruistic' Behavior in Rhesus Monkeys." *American Journal of
Psychiatry* 121: 584–85.

WHAT A DIFFERENCE A BAG MAKES
Riley, J. (March 17, 1967). "Saga of the Barefoot Bag on Campus."
Life: 72A–72B.

Zajonc, R. B. (1968). "Attitudinal Effects of Mere Exposure." *Journal of
Personality and Social Psychology,* Monograph Supplement 9 (2, Part
2): 1–27.

BEHIND THE WHEEL
Baron, R. A. (1976). "The Reduction of Human Aggressions: A Field
Study of the Influence of Incompatible Reactions." *Journal of
Applied Social Psychology* 6 (3): 260–74.

Deaux, K. K. (1971). "Honking at the Intersection: A Replication and
Extension." *Journal of Social Psychology* 84: 159–60.

Turner, C. W., J. F. Layton, & L. S. Simons (1975). "Naturalistic
Studies of Aggressive Behavior: Aggressive Stimuli, Victim
Visibility, and Horn Honking." *Journal of Personality and Social
Psychology* 31 (6): 1098–1107.

THE STANFORD PRISON EXPERIMENT

Haney, C., C. Banks, & P. Zimbardo (1973). "A Study of Prisoners and Guards in a Simulated Prison." *Naval Research Reviews* 30: 4–17.

Zimbardo, P. G. (2004). "A Situationist Perspective on the Psychology of Evil: Understanding How Good People Are Transformed into Perpetrators." In A. G. Miller, ed., *The Social Psychology of Good and Evil*. New York: Guilford Press. 21–50.

Zimbardo, P. G., C. Maslach, & C. Haney (1999). "Reflections on the Stanford Prison Experiment: Genesis, Transformations, Consequences." In T. Blass, ed., *Obedience to Authority: Current Perspectives on the Milgram Paradigm*. Mahwah, NJ: Erlbaum. 193–237.

THE UNRESPONSIVE BYSTANDER

MacCoun, R. J., & N. L. Kerr (1987). "Suspicion in the Psychological Laboratory: Kelman's Prophecy Revisited." *American Psychologist* 42: 199.

Schwartz, S. H., & A. Gottlieb (1976). "Bystander Reactions to a Violent Theft: Crime in Jerusalem." *Journal of Personality and Social Psychology* 34 (6): 1188–99.

Shotland, R. L., & W. D. Heinold (1985). "Bystander Response to Arterial Bleeding: Helping Skills, the Decision-Making Process, and Differentiating the Helping Response." *Journal of Personality and Social Psychology* 49 (2): 347–56.

Shotland, R. L., & M. K. Straw (1976). "Bystander Response to an Assault: When a Man Attacks a Woman." *Journal of Personality and Social Psychology* 34 (5): 990–99.

Ten: The End

FEAR FACTOR

Korn, J. H. (1997). *Illusions of Reality: A History of Deception in Social Psychology*. New York: State University of New York Press. 62–66.

HEARTBEAT AT DEATH

"Heart at Death" (November 14, 1938). *Life:* 20.

DYING ON ACID

Alsop, S. (1974). "The Right to Die with Dignity." *Good Housekeeping* 179 (2): 69, 130, 132.

Cohen, S. (1965). "LSD and the Anguish of Dying." *Harper's Magazine* 231: 69–78.

Phifer, B. (1977). "A Review of the Research and Theological Implications of the Use of Psychedelic Drugs with Terminal Cancer Patients. *Journal of Drug Issues* 7 (3): 287–92.

A SOUL IN THE BALANCE

Carrington, H. (1908). *The Coming Science*. Boston: Small, Maynard & Company.

MacDougall, D. (July, 1907). "Hypothesis Concerning Soul Substance." *American Medicine*, New Series, 2 (7): 395–97.

"Plan to Weigh Souls: Physician Proposes Experiment with Death Chair" (March 12, 1907). *Washington Post:* 3.

Roach, M. (2005). *Spook: Science Tackles the Afterlife*. New York: W. W. Norton & Company.

"Soul Has Weight, Physician Thinks" (March 11, 1907). *New York Times:* 5.

"Weight of the Soul: Experiments Made with Dying Men Arouse Sharp Comment" (August 3, 1910). *Washington Post* (reprinted from the *Lancet*): 2.

THE DAY THE WORLD DIDN'T END

"Chicago Unworried by Dire Prediction of Flood Tomorrow" (December 20, 1954). *Mexia Daily News:* 1.

Dawson, L. L. (1999). "When Prophecy Fails and Faith Persists: A Theoretical Overview." *Nova Religio* 3: 60–82.

THE LAST SURVIVOR

Berenbaum, M. (2001). "Rad Roaches." *American Entomologist* 47 (3): 132–33.

Sullivan, R. L., & D. S. Grosch (1953). "The radiation tolerance of an adult wasp." *Nucleonics* 11 (3): 21–23.

Visit **www.panmacmillan.com** to read more about all our books and to buy them. You will also find features, author interviews and news of any author events, and you can sign up for e-newsletters so that you're always first to hear about our new releases.

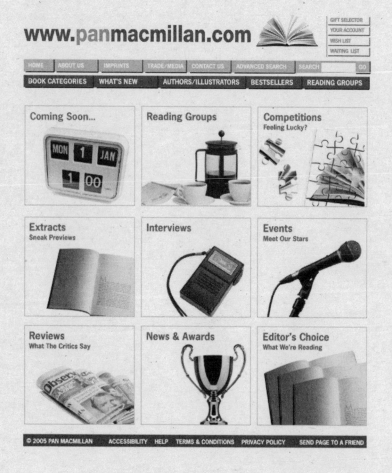